An Interdisciplinary Approach to the Theory and Practice of Wildlife Corridors

An Interdisciplinary Approach to the Theory and Practice of Wildlife Corridors

Conservation, Compassion, and Connectivity

By
Amy D. Propen

ANTHEM PRESS

Anthem Press
An imprint of Wimbledon Publishing Company
www.anthempress.com

This edition first published in UK and USA 2026
by ANTHEM PRESS
75–76 Blackfriars Road, London SE1 8HA, UK
or PO Box 9779, London SW19 7ZG, UK
and
244 Madison Ave #116, New York, NY 10016, USA

First published in the UK and USA by Anthem Press in 2024

© 2026 Amy D. Propen

The author asserts the moral right to be identified as the author of this work.

All rights reserved. Without limiting the rights under copyright reserved above, no part of this publication may be reproduced, stored or introduced into a retrieval system, or transmitted, in any form or by any means (electronic, mechanical, photocopying, recording or otherwise), without the prior written permission of both the copyright owner and the above publisher of this book.

British Library Cataloguing-in-Publication Data
A catalogue record for this book is available from the British Library.

Library of Congress Cataloging-in-Publication Data: 2025947470
A catalog record for this book has been requested.

ISBN-13: 978-1-83999-837-9 (Pbk)
ISBN-10: 1-83999-837-7 (Pbk)

Cover Credit: nazar_ab/istockphoto.com

This title is also available as an e-book.

CONTENTS

List of Figures   vii

Acknowledgments   ix

1. Introduction   1
2. Key Concepts and Ideas: Connectivity from a Species Perspective   13
3. Designing and Managing Wildlife Corridors   29
4. Emerging Theoretical Perspectives: Compassionate Conservation, Empathy, and Traditional Ecological Knowledge   43
5. The Wildlife Crossing on the Flathead Indian Reservation in Montana, USA: Respecting the Spirit of Place   57
6. The Monkton Wildlife Crossing and the Blue-Spotted Salamander: Vermont's First Amphibian Crossing Tunnels   69
7. The Railway from Oxford to London Marylebone: Transportation Upgrade Meets Compassion for Vulnerable Habitats   77
8. Aerial Corridors in Urban Environments: Light Pollution and Migratory Birds   83
9. The Papahānaumokuākea Marine National Monument: Traditional Ecological Knowledge and Marine Protected Areas   91
10. Looking Ahead: New Perspectives and Best Practices Related to Wildlife Corridors   99

Bibliography   111

Index   119

LIST OF FIGURES

2.1:	Habitat fragmentation	14
2.2:	A breach in corral-style fencing, likely created by deer passing through this residential area, was intentionally left as is by the landowner to allow deer and other wildlife to more easily exit the property	19
2.3:	Photograph of a wildlife underpass crossing culvert for animals under a highway in the Netherlands	19
2.4:	Photograph of an overpass over the Trans-Canada highway designed for wildlife migration/crossing	20
2.5:	Aerial view of roadside vegetation	21
2.6:	Photograph overlooking Boston Common, Charles River, and Back Bay: part of the Emerald Necklace System	22
5.1:	Map: Locations of wildlife crossing structures along U.S. Hwy 93, Montana, USA	59
5.2:	Map: Locations of wildlife exclusion fences along U.S. Hwy 93, Montana, USA	60
5.3:	Wildlife overpass, or "Animals' Bridge," U.S. Hwy 93, Flathead Indian Reservation, Montana, USA	61
5.4:	Wildlife jump-out or escape ramp along U.S. Hwy 93, Montana, USA	61
6.1:	Blue-spotted salamander	70
6.2:	Monkton wildlife crossing tunnel	73
9.1:	Laysan Albatross colony on Papahānaumokuākea Marine National Monument, Midway Island, Midway Atoll, Hawaiian Islands	92

ACKNOWLEDGMENTS

This book is devoted specifically to the study of wildlife corridors, but really, at its core, it is about coexistence. That is, through its focus on the theory and practice of wildlife corridors and corridor ecology, this book considers how people and wildlife can better coexist in a time of increasing habitat fragmentation due largely to the impacts of human development.

We humans have much to offer and much to learn from our nonhuman kin, and in many ways, the goal of reconnecting fragmented habitats and ecosystems represents one way in which we may strive to do both. By better conceptualizing and putting into practice wildlife corridors, we can make accessible once-lost or almost-lost habitats for our community of wildlife, and we can, in the process, promote a broader culture of connectivity and coexistence that has the potential to lead toward more sustainable futures.

This project has benefited from the input and feedback of those whose work pertains to wildlife corridors and road ecology in various ways.

Thank you to the Western Transportation Institute at Montana State University (WTI-MSU) for permission to reprint the maps in Figures 5.1 and 5.2. Thank you to the Vermont Agency of Transportation for permission to reprint the photos in Figures 6.1 and 6.2. Thanks to Chris Slesar, environmental resources coordinator for the State of Vermont Agency of Transportation. Thank you to Marcel Huijser, Ph.D., research ecologist at Montana State University, and talented wildlife photographer, whose photographs are featured in Figures 5.3 and 5.4.

I appreciate the support and assistance of Jebaslin Hephzibah at Anthem Press, as well as the support of series editor Lawrence Susskind at MIT not only for taking an interest in the project early on but also for his patience while I completed it on a slightly longer timeline than originally intended. Thank you also to the reviewers for their comments and feedback along the way.

Thank you to Mary Schuster for her encouragement and good conversation over the course of this project. Finally, thank you to my parents and wildlife advocates, Beverly and Michael Propen, whose support and interest in this book helped fuel my own interest in writing it.

Chapter 1

INTRODUCTION

I grew up in Orange, Connecticut—a fairly rural suburb in southern Connecticut that is home to about 14,000 people and encompasses about 17 square miles. Orange is about two hours north of New York City and about five miles inland from the Atlantic coast and the Long Island Sound. The Long Island Sound watershed includes many cities and towns along the New York and Connecticut coastlines. Countless rivers and streams flow through these small towns and eventually discharge into the Sound, making for scenic drives and rest stops along the way. But the Wepawaug River, Indian River, and Oyster River, all of which flow through the town, do far more than serve as scenic rest stops, property markers, and fishing spots for local residents and visitors. A look at the popular *eBird* app, for instance, reveals 118 songbird, seabird, and waterfowl sightings along the Oyster River alone, which runs through Orange and empties into Long Island Sound.[1] Species including Ring-billed Gulls, Great Egrets, Piping Plovers, and Belted Kingfishers depend on these streams and rivers for sustenance and safe habitat.

The house I grew up in is just up the street from a section of the Indian River that still runs behind some of the homes in the neighborhood. In high school, I'd often walk down the street and hike into the wooded area along that small portion of the Indian River. There, I'd sometimes come across White-tailed Deer drinking from the river as they would pass through the neighborhood and box turtles who lived along the stream but who would, for some reason, often lay their eggs closer to the street. At the time, I did not have the language to describe the Indian River or the Long Island Sound as "ecologically significant areas."[2] Nor did I think of these local streams and rivers as "wildlife corridors," per se; however, it is clear now that, as I watched those deer, turtles, and local foxes traverse the Indian River and make their way through portions of neighbors' fencing to continue their travels through local green spaces and state forest areas, that they were creating their own corridors and paths of connectivity through now-developed areas that offer only partially contiguous landscapes. In fact, these species have existed in this region long before I grew up there and likely before the town itself was settled in 1822. That is, what is now Orange, Connecticut, was originally inhabited by the Paugusset and Algonquian peoples before it was eventually taken over by English colonists; throughout the 1800s, the town continued to suburbanize, and by the early 1900s, railroad and highway systems cut through the region.[3] As human development steadily pushed its way through previously untouched ecosystems, species like deer, fox, turtles, waterfowl, and bears had to get more creative in navigating their way through once-uninterrupted landscapes. And today, as they try to safely make their way through areas of increasingly significant human use, we must ask ourselves how we might account for and mitigate the human development that has fragmented these ecosystems and posed challenges to vulnerable species who, as they attempt to make their way through these landscapes, encounter people more and more frequently. In other words, how might we best coexist with these vulnerable, nonhuman species who are, in fact, also our neighbors?

While my aforementioned example describes smaller, less well-defined, or "official" wildlife corridors, wildlife corridors can be defined in different ways depending on the context, may occur naturally or be human-made, and can happen at different levels of scale. In this book, I understand wildlife corridors as spatial, infrastructural interventions that attempt to reconnect fragmented habitats and reestablish or protect biodiversity through necessarily interdisciplinary systems of planning, design, and knowledge-making. Moreover, I argue that if designed with the principles of compassion and empathy in mind and if efforts are made to incorporate the voices of multiple stakeholders, including Indigenous perspectives, wildlife corridors have the potential to perform conservation practice grounded in an ethic of care and empathy, as well as to reshape how we understand what counts as "natural" places and our ideas about "who belongs where." That is, we can no longer perceive national parks and protected spaces as intended only for wildlife, just as we can no longer perceive urban spaces as intended only for people.[4] As we design cities, "we are designing cultures and communities."[5] And as we negotiate how to account for climate change in our design and living processes, we must learn to "live symbiotically with nature."[6]

On the whole, wildlife corridors "provide continuous habitat for species to move on their own, [and] are a reasonable and effective means for ensuring connectivity in the landscape."[7] From small gaps in residential fencing that help prevent wildlife from getting trapped in residential yards to larger freeway overpasses meant to help species cross highways safely, wildlife corridors can be conceptualized, planned, and carried out in a range of ways, depending on the purposes, contexts, landscapes, scale, and stakeholders involved. However, the goal of achieving and maintaining ecological connectivity, or the "unimpeded movement of species and the flow of natural processes that sustain life on Earth," is common across contexts.[8]

This book charts some best practices and makes some new theoretical contributions related to the design and creation of wildlife corridors in Anthropocene times, or at a time when the impact of human development on the natural world can no longer be denied, that is, if it ever could. In doing so, this book aims to provide the foundational knowledge necessary for a general and credible understanding of connectivity projects but also endeavors to make a unique theoretical contribution to current knowledge about wildlife corridors by arguing that theories about compassionate conservation, entangled empathy, and traditional ecological knowledge (TEK) should inform the planning and creation of wildlife corridor projects.[9]

Briefly put, and described in more detail below and in later chapters, the concept of *compassionate conservation* suggests that the life of every individual species matters in decisions about wildlife conservation and policy.[10] The concept of *entangled empathy*, as theorized by philosopher Lori Gruen, relates to humans' ability to empathize with other animals. As Gruen describes, entangled empathy involves learning from the perspective of another's experience and "being able to understand what another being feels, sees, and thinks, and to understand what they might need or desire."[11] *Traditional ecological knowledge (TEK)*, as articulated by Fikret Berkes, understands knowledge about the environment and ecosystems as cumulative, conveyed over time and through generations by cultural transmission, and focused on "the relationship of living beings (including humans) with one another and with their environment."[12] In this book, I do my best to describe these theories in a manner that is accessible to various audiences, such that this work will be accessible across audiences, including to scholars of conservation biology, environmental studies, feminist geographies, and other related disciplines, as well as to community planners, environmentalists, and any individuals interested in coexisting peacefully with our nonhuman kin.

In this chapter, I provide some necessary context and terminology and outline some of the book's key arguments. As landscapes become fragmented due to both human-induced and natural causes, the migration and movement of vulnerable species across those landscapes becomes more difficult, if not impossible. While landscape fragmentation can happen due to human intervention or more naturally, this book tends to focus more so on fragmentation wrought by the products of human development, including linear transport infrastructure (LTI) such as roads, railways, and canals, and their impacts on habitat fragmentation and biodiversity loss.[13] Roadways are perhaps one of the greatest ongoing sources of habitat fragmentation; moreover, the human activity and sprawl that they enable put at risk the long-term survival of many species.[14] I argue that the creation of wildlife corridors can help work against such fragmentation by maintaining or restoring connectivity "between species, ecosystems, and ecological processes" at various scales.[15] In order to achieve ecological and landscape connectivity in a way that also aligns with the needs of vulnerable species, this book also suggests that wildlife corridors should be conceptualized from a mindset of compassionate conservation and should account for TEK whenever it is relevant and feasible to do so.

Connectivity projects such as wildlife corridors are necessary because, when land is set aside or used for human activities, "habitats that were once contiguous become divided into separate fragments."[16] If species are unable to move between these fragmented areas, they become at risk for inbreeding or extinction.[17] Wildlife corridors attempt to restore connectivity and create "bands of forest habitat that are large and intact enough that they provide animals with an important bridge between larger blocks of habitat."[18] Providing such linkages between habitats reduces these risks, helps reduce fragmentation, and maintains genetic biodiversity and a population's health by allowing wildlife to move more freely in and out of different areas.

This book more explicitly charts some best practices and unpacks some theories related to wildlife corridors that take into account issues like public communication, environmental advocacy, and animal studies, as well as some less salient perspectives, such as approaches to biodiversity that draw on theories of compassionate conservation, entangled empathy, and TEK.

Wildlife corridor projects are necessarily interdisciplinary undertakings; they require both input from and confidence on the part of multiple stakeholders, such as biologists, state and tribal agencies, nonprofit organizations, local grassroots and public groups, private landholders, and others. The land that comprises a single corridor system is often owned by several groups or private landowners and thus requires an inclusive management approach that can account for the needs of many parties, including the wildlife that the corridors are designed to protect in the first place. That said, compassionate conservation and TEK would advocate for coexistence among people *and* wildlife, and many corridor projects have likewise demonstrated the most success when they can function in the best interests of people and wildlife—not always an easy goal to achieve. The relative success of corridor projects is thus notoriously difficult to gauge, in part because every project is context-specific and in part because "success" is often defined differently by different groups or viewed at more micro-levels in terms of the migration of a specific species or the restoration of a specific type of flora.[19] This book is mindful of these challenges as it nonetheless argues for a holistic approach to wildlife corridors that attempts to account as much as possible for a broad and varied range of stakeholder voices, including those of the vulnerable nonhuman species that underpin the need for corridor projects in the first place. In this way, my hope is that this book contributes to the ongoing dialogue about corridors by advocating for an approach that integrates ideas about compassionate conservation, entangled empathy, and TEK to the extent possible in any given corridor project. That is, one of this book's key contributions is its addition of new theoretical perspectives applied to the study of wildlife corridors, namely an interdisciplinary perspective that is informed by compassionate conservation, entangled empathy, and ideas about TEK. These perspectives can all shape more nuanced understandings of connectivity projects in an age of climate change, or a climate crisis.

Compassionate Conservation

Broadly speaking, an approach informed by *compassionate conservation* would advocate for viewing species as individuals, and it would thus take individual species into consideration within conservation policy.[20] Compassionate conservation is, at its core, grounded in an ethic of "do no harm." A compassionate conservation approach, as ecologist and evolutionary biologist Marc Bekoff describes, would not advocate for a conservation project that would cause harm to any species in the process of reconnecting habitats, for instance.[21] Part of the goal of compassionate conservation is to prompt and foster a critical awareness of how we understand our relationships with wildlife and to foster an ethic of peaceful coexistence in which decision-making about conservation practice and policy is grounded in a consideration of the most compassionate choice for *all* beings. Of course, this is easier said than done, and we may easily complicate the notion that conservation is still, at the end of the day, about humans making decisions on behalf of wildlife. I argue, however, following Bekoff's groundbreaking work on compassionate conservation, that when we begin to ask the right questions, or questions that are more critically engaged and that come from a place of compassion and kinship and less from the position of attempting to regulate or control our environment in a top-down fashion or valuing certain kinds of species over others, we might begin to acknowledge the rich, multispecies entanglements that can inform a compassionate conservation ethic.

Entangled Empathy

Moreover, a compassionate conservation ethic is not incompatible with ideas about what philosopher Lori Gruen refers to as entangled empathy. In distinguishing between sympathy and empathy, Gruen writes that "sympathy involves maintaining one's own attitudes and adding them to a concern for another. Sympathy for another is felt from the outside, the third-person perspective. [...] Empathy, however, recognizes connection with and understanding of the circumstances of the other."[22] Gruen acknowledges that this understanding may be incomplete and "often is in need of revisions. However, the goal is to try to take in as much about another's situation and perspective as possible."[23] Gruen has written extensively on the topic of how we empathize with nonhuman animals and has developed a concept that she calls "entangled empathy." Entangled empathy, or "being able to understand what another being feels, sees, and thinks, and to understand what they might need or desire," she says, "requires a fairly complex set of cognitive skills and emotional attunement."[24] Because entangled empathy focuses on "another's experiential well-being," this kind of empathy tends to lead to action based on an assessment of what seems to be the best, or most compassionate choice, in helping to pursue the well-being of another.[25] Such decisions are, nonetheless, often fraught and not necessarily clearly defined. Based on these ideas, though, it becomes clear that both compassionate conservation and entangled empathy are very much aligned with the philosophies, discourses, and practices underpinning wildlife conservation and may be applied to wildlife corridor projects.

Traditional Ecological Knowledge (TEK)

Finally, both compassionate conservation and entangled empathy are compatible with ideas about TEK. While there are varying and nuanced definitions of TEK, this book aligns with the working definition put forth by Fikret Berkes, which understands TEK as "a cumulative body of knowledge, practice, and belief, evolving by adaptive processes and handed down through generations by cultural transmission, about the relationship of living beings (including humans) with one another and with their environment."[26] In this sense, TEK, with its roots in anthropology, refers to both *ways of knowing*, as in ways of *being in the world*, and information and knowledge *about things and processes*.[27] Such distinctions are important for wildlife corridor projects in terms of how TEK is leveraged and incorporated in the design and planning processes.

Moreover, as Dan Shilling describes, we can look to a few common threads that inform overarching ideas about TEK: namely, the idea that "reciprocity and respect" define the connections between "all members of the land family"; the idea that a deep respect for nature should inform all activities related to the natural world; and that our relationship to land should be informed by "something other than economic profit"; likewise, ideas about land ownership must be viewed with a critical awareness that moves beyond hierarchical ideas that privilege the human role in ecosystems above all else; and "each generation has a responsibility to leave a healthy world to future generations."[28] Indigenous studies scholar, biologist, and botanist Robin Wall Kimmerer helps illuminate these ideas by further describing the ways that gratitude and a culture of reciprocity should inform our relationship with the land and ecosystems; she writes that, looking through "the lens of traditional Indigenous philosophy the living world is understood, not as a collection of exploitable resources, but as a set of relationships and responsibilities."[29] Along similar lines, Ament et al. note that: "Full, effective and genuine participation of local communities and Indigenous peoples is necessary and increases the potential for LTI projects to benefit all stakeholders."[30] In this book, I consider how wildlife corridors may function in sync with these ideas and philosophies and how such work can benefit the ecosystems and vulnerable species that inhabit them, in the process.

Wildlife corridors illustrate the challenges of compassionate coexistence between wildlife and people, in that they entail a spatial reorganization of natural and urban places, which may impact a wide range of stakeholders through an effort to preserve biodiversity and promote a culture of coexistence among human and nonhuman species. Through the connectivity that they help foster, they also challenge more boundaried, human-centric assumptions about what counts as "nature," or "natural" versus "urban" places—boundaries that also begin to dissolve when understanding Indigenous approaches to human/more-than-human animal relationships. Literary scholar Jeffrey Jerome Cohen acknowledges that "'Nature' is a difficult word. It names something at once 'everywhere and nowhere,' leading some critics to argue that we are better off without the term."[31] Cohen then argues that to better access our sense of ethical connectedness to the world, we must understand "nature" in more nuanced ways—as always mediated by humans, and as intersecting multiple kinds of bodies and structures, including humans and other animals and species across ecosystems and places; in these ways, then, nature also incorporates what we commonly think of as "wildlife." Likewise, this book understands "wildlife" from a more relational perspective. Anthropocentric, or human-focused understandings of wildlife, tend to picture animals as "over there," in places separate from humans. But as feminist philosopher Val Plumwood explains, "'Wilderness' is not a place where there is no interaction between self and other, but one where self does not impose itself."[32] As geographer Jamie Lorimer similarly describes, wildlife lives among us and includes even those "feral plants and animals that inhabit urban ecologies."[33] Thus, when I use terms such as "nature" and "wildlife" throughout this book, I use them with the understanding that they are aligned with these broader, more relational, and less anthropocentric perspectives.

At a time when we face even more biodiversity and species loss than ever before, it becomes more necessary than ever to look for, engage with, and draw upon more nuanced perspectives that are sensitive to the needs of wildlife on levels perhaps different from but nonetheless compatible with more mainstream conservation science. That is, as the U.S. Fish & Wildlife Service notes: "An increasing number of scientists and Native people believe that Western Science and TEK are complementary. Although an integration of indigenous and western scientific ways of knowing and managing wildlife can be difficult to achieve, successful integrations have occurred," and the merging of TEK and western scientific perspectives is becoming more common in conservation projects.[34]

This book draws upon some of the foundational works in corridor ecology that theorize the design, planning, and implementation of wildlife corridors, while also building on and extending those ideas to argue for the inclusion of compassionate conservation, empathy, and TEK in wildlife corridor projects. For additional exploration of corridor ecology, see, in particular, Hilty et al. (2019, 2020) and Anderson and Jenkins (2006). Finally, I should note that this book proceeds from the understanding that there is likely no perfect solution in the ongoing efforts to reconnect fragmented habitats, especially in an age of climate crisis and human development; however, that does not mean we ought not to try. While it may be unrealistic to assume that the perfect, tidy corridor project exists, I suggest that an approach informed by compassion and TEK, and an interest in coexistence among all beings, nonhuman and human, can help forge a productive path forward for corridor projects across a range of contexts. In doing so, I hope to extend the conversation and generate new ideas about inclusive practices for how we might best learn to coexist with wildlife in an age of climate change and unabating human development, or in the age of the Anthropocene. With these arguments, contexts, and ideas in mind, this book proceeds along the lines described in the chapter summaries below.

Chapter Summaries

Following this Introduction, Chapter 2 continues to provide a context and framework for understanding wildlife corridors by focusing on key concepts and ideas related to wildlife corridors and corridor ecology. The chapter addresses different ways of understanding *connectivity*; the concepts of *habitat fragmentation* and *connectivity*; and the importance of *scale*, such as larger- or smaller-scale corridors, in considering the ways that a corridor project may influence the *biodiversity* of a given region and ecosystem. It describes the main types of corridors, such as natural, linear linkages versus artificial structures and green infrastructure projects like tunnels and overpasses. It then discusses the range of possible functions of such corridors; for instance, some projects are longer-term and involve entire landscapes, with the goal of making multidirectional connections and restoring entire ecosystems, whereas other projects might be shorter-term in scope, with the goal of linking two blocks of habitat to restore a specific target species. Corridors may also be implemented in areas of varying conservation status, such as areas of unlogged forest, regenerated areas, planted areas such as windbreaks or greenways, and so on. In each of these instances, the various social and economic contexts, types of species, and local ecologies must be accounted for in the goals and approaches of the corridor project.[35]

Chapter 3 focuses on the design and management of wildlife corridors. The chapter describes the need for a "biodiversity vision," or the larger goals associated with any connectivity project, and how such goals should map to the design process. It discusses the factors upon which landscape connectivity depends, such as understanding gaps in the habitat or the existence of other natural pathways that should be protected or leveraged. It provides some general design guidelines while also keeping in mind that every project is unique in its context; nonetheless, corridor design should strive to leverage existing natural corridors, mitigate for climate change, minimize the connection of disturbed habitats with less-disturbed areas, and build in redundancies.[36] Moreover, design should account for local context and knowledge, as well as identify competing objectives or strategic visions. These latter points also begin to illustrate the challenges of wildlife corridors as they pertain to the various stakeholders and perspectives associated with such projects. With these ideas in mind, this chapter addresses the need to engage a wide range of stakeholders in corridor projects; the need to balance biodiversity issues with social and cultural contexts; the challenges of understanding and planning for budgetary issues and identifying various costs associated with corridor projects and their maintenance; the challenges of building a support base and educating the public about the needs for such projects; and the various rationales for public support and sustained governance, which are often rooted in both intrinsic and extrinsic valuing of wildlife and ecosystems. Following this discussion of the various rationales for how we value wildlife, Chapter 4 then explores some of the philosophies that the book argues should underpin any corridor project.

Chapter 4 explores some of the emerging theoretical perspectives that this book argues should underpin and complement wildlife corridor projects. The chapter adds to the ongoing dialogues about wildlife corridors by describing how wildlife corridor projects may benefit from the inclusion of perspectives grounded in compassionate conservation, entangled empathy, and TEK.

Subsequently, I argue for an approach to connectivity and coexistence that rethinks more commonly perceived boundaries and hierarchies in human relationships with wildlife, that works against anthropocentric and hierarchical thinking, and that does not necessarily privilege one kind of species, such as charismatic megafauna, over another. Ultimately, the chapter argues that we must strive to design spaces that allow people and wildlife to *coexist* in the Anthropocene. From here, the book moves into a set of illustrative cases that help show the various ways that we can incorporate connectivity at a range of scales, through examples of both larger- and smaller-scale projects that can help promote a culture of connectivity and coexistence.

Chapters 5 through 9 are then comprised of five illustrative cases that demonstrate ideas and theories about wildlife corridors in practice. The goal of these cases is to help further contextualize some of the theories and philosophies discussed throughout the book. These cases are not necessarily organized based on geographic location, per se; rather, I have chosen these cases for their ability to help illustrate connectivity projects across a range of ecosystems and scales, to highlight efforts to protect a range of species and areas of biodiversity, and to help illustrate, to varying extents, the philosophies that this book suggests should underpin corridor design.

These five cases involve: the wildlife crossing on the Flathead Indian Reservation in Montana, which incorporated TEK in a large-scale highway reconstruction project; the Monkton Wildlife Crossing, which is the state of Vermont's first amphibian crossing tunnel; the railway from Oxford to London Marylebone in the UK, which involves the interplay of transportation and vulnerable ecosystems; the somewhat newer concept of "aero-corridors" in U.S. cities and their implications namely for light pollution and migratory birds; and the Papahānaumokuākea Marine National Monument in Hawaii, which integrated TEK in its design and planning. These illustrative cases each speak in different ways to the concepts addressed in the earlier chapters. Some illustrate very clear examples of integrating TEK, for instance, while others focus more on ideas about citizen outreach and education, grassroots efforts, and achieving a balance between management and compassionate conservation.

Finally, Chapter 10 considers some best practices in light of these new perspectives related to the planning, design, and implementation of wildlife corridors. In this brief section, I conclude the book by bringing together the concepts, philosophies, and case examples to consider some best practices and new theoretical directions for ongoing planning and development of wildlife corridors and connectivity projects. It is my hope that, taken together, the ideas and examples in this book can contribute to and extend the dialogue about wildlife corridors in new and productive ways.

Notes

1. eBird, "Oyster River."
2. CT DEEP, "Long Island Sound Blue Plan."
3. Orange Historical Society, "History."
4. Pratt, "How a Lonely Cougar."
5. Watson, *Lo—TEK*, 399.
6. Watson, *Lo—TEK*, 399.
7. Conservation Corridor, "IUCN Guidelines."
8. Ament et al., *Addressing Ecological Connectivity in the Development of Roads, Railways and Canals*, xii.
9. On compassionate conservation, see Bekoff, *The Animal Manifesto*; Bekoff, *Rewilding Our Hearts*; on entangled empathy, see Gruen, *Entangled Empathy*; on traditional ecological knowledge, see Berkes, *Sacred Ecology*; Watson, *Lo—TEK*.
10. See Bekoff, *The Animal Manifesto*; Bekoff, *Rewilding Our Hearts*.
11. Gruen, *Entangled Empathy*, 50.
12. Berkes, *Sacred Ecology*, 7.
13. Ament et al., *Addressing Ecological Connectivity in the Development of Roads, Railways and Canals*, viii.
14. Beckmann and Hilty, "Connecting Wildlife Populations in Fractured Landscapes," 9–10.
15. Anderson and Jenkins, *Applying Nature's Design*, 4.
16. Vermont Natural Resources Council, "Wildlife Corridor Protection."
17. VNRC, "Wildlife Corridor Protection."
18. Stowe Land Trust, "Putting."
19. See Anderson and Jenkins, *Applying Nature's Design*; Hilty et al., *Corridor Ecology*.
20. See Bekoff, *The Animal Manifesto*; Bekoff, *Rewilding Our Hearts*.
21. MacKay, "Where Compassionate Conservation."
22. Gruen, *Entangled Empathy*, 44–45.
23. Gruen, *Entangled Empathy*, 45.
24. Gruen, *Entangled Empathy*, 50.
25. Gruen, *Entangled Empathy*, 51.
26. Berkes, *Sacred Ecology*, 7.
27. Berkes, *Sacred Ecology*, 8.
28. Shilling, *Traditional Ecological Knowledge*, 12.
29. Kimmerer, *Braiding Sweetgrass*, 27.
30. Ament et al., *Addressing Ecological Connectivity in the Development of Roads, Railways and Canals*, ix.
31. Cohen, *Stone*, 12.
32. Plumwood, *Feminism and the Mastery of Nature*, 164.
33. Lorimer, *Wildlife in the Anthropocene*, 7.
34. US Fish & Wildlife, "Traditional Ecological Knowledge."
35. See Anderson and Jenkins, *Applying Nature's Design*; Hilty et. al, *Corridor Ecology*.
36. See Anderson and Jenkins, *Applying Nature's Design*; Hilty et al., *Corridor Ecology*.

Chapter 2

KEY CONCEPTS AND IDEAS

Connectivity from a Species Perspective

To suggest that wildlife corridors are a necessary component of restoring lost habitat connectivity requires an understanding of some of the concepts that shaped the conceptualization and implementation of wildlife corridors in the first place. To begin, as defined by the research group *Conservation Corridor*, a corridor is "a habitat whose main function is to connect isolated patches of habitat that would otherwise be inaccessible."[1] Next, understanding the value of wildlife corridors also requires knowledge of the concept of connectivity or the ease with which species are able to move between different appropriate habitats.

Understanding Connectivity

As mentioned in the Introduction, ecological connectivity refers to "the unimpeded movement of species and the flow of natural processes that sustain life on Earth."[2] As habitats become increasingly fragmented, and as biodiversity continues to be threatened by the processes of infrastructure development, climate change, and other forces driving the Anthropocene, it becomes even more critical to protect connectivity (Figure 2.1). As Hilty et al. note in their 2020 IUCN report, *Guidelines for Conserving Connectivity through Ecological Networks and Corridors*: "Science overwhelmingly shows that interconnected protected areas and other areas for biological diversity conservation are much more effective than disconnected areas in human-dominated systems, especially in the face of climate change."[3] Moreover, to achieve this, the 2020 IUCN report recommends establishing an *ecological network for conservation* to help establish ecological corridors and achieve connectivity. The IUCN defines an *ecological network for conservation* as "a system of core habitats [...] connected by ecological corridors, which is established, restored as needed and maintained to conserve biological diversity in systems that have been fragmented."[4] This network, or system of core habitats, would ideally be

Figure 2.1 Habitat fragmentation. Credit: Wirestock.

comprised of protected areas, other effective area-based conservation measures (OECMs), and other intact natural areas. Together, these areas would constitute an ecological corridor; in other words, functioning in the positive, an ecological network would be connected with ecological corridors.[5] An ecological corridor "is a clearly defined geographical space that is governed and managed over the long term to maintain or restore effective ecological connectivity."[6] While it is possible to achieve connectivity conservation goals without the presence of a network per se, ecological networks for conservation "are more effective in achieving biodiversity conservation objectives than a disconnected collection of individual protected areas and OECMs because they connect populations, maintain ecosystem functioning and are more resilient to climate change."[7] The idea of "connection," in the sense being discussed here, refers to "the enabling of movement by individuals, genes, gametes and/or propagules."[8] While connectivity can happen without the establishment of wildlife corridors, corridors are emerging as one of the best ways to achieve and enhance connectivity within landscapes and for the wildlife who inhabit them.[9]

As Anderson and Jenkins note, three main factors influence a corridor's connectivity: "the number and size of gaps in the corridor habitat, the presence of alternative pathways or networks, and the existence of larger habitat patches, or 'nodes,' along the corridor."[10] A corridor habitat may include smaller stepping stones, for instance, which are smaller "patches" of habitat that afford wildlife room to move "through the landscape without a continuous corridor, but they require individuals to traverse through lower quality habitat in between."[11] In this way, a patch is a central area of habitat—sometimes a smaller corridor, that is separate from the surrounding matrix, and a matrix is the landscape that surrounds a corridor.

In marine environments, as the illustrative case regarding Hawaii's Papahānaumokuākea Marine National Monument will discuss, ecological corridors can facilitate movement by native animals and plants. Marine ecological corridors can also work in conjunction with marine protected areas (MPAs) to help protect species from the human impacts of shipping lanes and acoustic testing in the ocean. Marine corridors can be integral for species that use "different environments at different stages of their life cycles. For example, marine turtles nest on beaches and may use coastal waters before moving into the high seas, while certain fish may need to migrate to reach a spawning aggregation site."[12] Additionally, marine ecological corridors can vary in scale; that is, they "may need to be quite large given the extent to which oceanic currents, eddies and tides affect processes and the recruitment of organisms," or they may be smaller-scale "to protect migrations of a few kilometres, such as those of red crabs (Gecardoidea natalis) on Australia's Christmas Island."[13]

If wildlife is challenged to move through a landscape on their own due to habitat fragmentation, conservation biology projects may also undertake what is referred to as "assisted migration," in which humans physically move or relocate individual species of wildlife to a new, better location that might not otherwise be reachable. Assisted migration can potentially help populations in certain instances but can be controversial as a management tool, depending on the species and habitat in question. Corridors thus continue to emerge as the more "effective means for ensuring connectivity" both from the perspective of landscape and species; that is, there are two main ways to understand the influence of connectivity: its influence on the landscape, and its influence on species—both are two sides of the same coin but are interconnected.[14]

Structural and Functional Connectivity

There are two main ways of defining and understanding the connectivity that corridors can provide: structurally and functionally. Structural connectivity refers to the physical characteristics of a habitat that make it permeable and thus able to connect to another section of the habitat. Subsequently, whether and how actual species make use of that structural connectivity refers to how the habitat actually functions, or its functional connectivity. Functional connectivity determines how species are affected by conservation management strategies. In other words, when it comes to wildlife corridors, structural connectivity refers to how a corridor is built and how the physical elements of the corridor interact with each other, whether land-based or sea-based. How the corridor serves the needs of the species in question refers to its actual function in the world or its functional connectivity. A wildlife corridor is considered to have "high functional connectivity if it supports dispersal and movement, and protects ecological processes."[15] Or, put differently, structural connectivity describes connectivity from a landscape or seascape perspective, while functional connectivity describes connectivity from the perspective of individual species.

It is possible to have structural connectivity with limited functional connectivity or functional connectivity with limited structural connectivity, and each is measured differently; for example, structural connectivity may be measured with landscape metrics offered by tools from geographic information science (GIS), while functional connectivity may be measured with data visualization and other tracking tools, which can also be related to GIS. Each metric is useful for different purposes; that is, landscape metrics are generally better at representing space, while tracking metrics are better for getting a picture of the real-time movements of species between spaces.[16]

If species are unable to move between fragmented landscapes, they become at-risk of inbreeding or extinction.[17] Wildlife corridors attempt to restore connectivity and improve species' *gene flow*, or "the transfer of genetic material through a population, which can be used as a measure of how connected populations are."[18] To maintain gene flow among species ultimately helps maintain biodiversity and a population's health.

Connectivity and Scale

A wildlife corridor's scale, such as a larger- or smaller-scale corridor, also influences and is influenced by the biodiversity of a given region and ecosystem. We may view wildlife corridors in terms of both their spatial and temporal scales. The spatial scale of a corridor may generally be understood at the local, regional, or continental or cross-continental level.[19] At the local level, a smaller underpass may allow particular species to get across a roadway system without entering the road or highway; in Monkton, Vermont, for instance, amphibian crossing tunnels enable thousands of salamanders and frogs to safely traverse road systems during their annual journey to and from their breeding pools.[20] At the regional level, a river system like the Saint Croix National Scenic Riverway, which spans Wisconsin and Minnesota, can provide a haven for wildlife such as native mussels, dragonflies, eagles, osprey, kingfisher, warblers, raccoons, foxes, and bears, among others.[21] Larger-scale corridor initiatives can help protect at-risk landscapes from fragmentation;

such initiatives often require cooperation across countries and continents and communication across stakeholders. The initiative called Two Countries, One Forest, for instance, is a collaborative organization made up of conservation groups, researchers, and citizens that spans the United States and Canada; this international group focuses on the "protection, conservation and restoration of forests and natural heritage from New York to Nova Scotia, across the Northern Appalachian Acadian ecoregion."[22]

Connectivity projects can also have functions and goals that differ temporally; for instance, some projects are longer-term and involve entire landscapes, with the goal of making multidirectional connections and restoring entire ecosystems, whereas other projects might be shorter-term or smaller-scale in scope, with the goal of linking two blocks of habitat to restore a specific target species, especially during particular times of year when that species must migrate to breed, as is the case with the Monkton corridor in Vermont.

The Benefits of Corridors for Nonhuman Species

The responses of different species to wildlife corridors are also scale-specific and differ based on the needs of that species. White-tailed deer, for instance, "are capable of exploiting large portions of a landscape and move through both narrow habitat corridors and the disturbed, surrounding matrix," or the landscape that surrounds the corridor.[23] In contrast, understory birds tend to be far less habitat-tolerant. They may use corridors linking similar forested habitats only under extreme conditions. How species respond to corridors can vary based on the scale of the habitat.[24] For this reason, Anderson and Jenkins argue that corridor research should be mindful of the connections between the particular species in question and the scale of the project.[25]

Meyer et al. likewise note that effective biodiversity conservation requires that corridors must meet the needs of multiple species.[26] To this end, Meyer et al. recently employed "a combination of data types and analytical approaches to identify and compare corridors for several large mammal species within the Panama portion of the Mesoamerican Biological Corridor."[27] The Mesoamerican Biological Corridor (MBC) is a region comprised of Belize, Guatemala, El Salvador, Honduras, Nicaragua, Costa Rica, Panama, and some southern states of Mexico. The region has long served as a natural land bridge from South America to North America; the Isthmus of Panama in particular "is the last and narrowest portion of the MBC which connects Mesoamerica to South America and has acted as an intercontinental land bridge for a large suite of taxa—including mammals—for millions of years."[28] Moreover, Meyer et al. note that identifying key areas that can facilitate movement and gene flow for mammals throughout Panama will in turn bolster ecosystem function and preserve biodiversity more broadly because so many species use the MBC for migratory purposes.[29] In their study of the MBC, Meyer et al. illustrate the importance of biological corridors not only "for connecting large habitat areas for seasonal migration" but also for protecting biodiversity by aiding in the dispersal of individuals across populations "to increase gene flow and long-term population viability."[30] Like Anderson and Jenkins, they note that it is becoming more widely recognized that knowledge of animal behavior "rather than expert opinion alone is of paramount importance" for understanding "environmental effects on functional connectivity."[31]

On the one hand, the interplay of structural and functional connectivity is important in how a corridor works. On the other hand, while functional and structural connectivity work together, it has become more widely accepted over the past two decades that functional connectivity better accounts for connectivity from a species' perspective.[32] I would add to this point by noting that functional connectivity implicitly aligns more with a compassionate conservation and entangled empathy perspective, since it seeks to understand and support connectivity from the perspective of individual species.

Types of Corridors

Corridors may vary in type and purpose and thus can provide different kinds of connectivity depending on the landscape as well as species' needs. Landscapes may inadvertently serve as corridors, even when they haven't been explicitly designed as such by humans. That is, naturally existing boundaries formed by a dense row of shrubs or low trees can serve as corridors for species, as can roadside vegetation. Individuals and private landowners can also create corridors or foster connectivity by making smaller-scale adjustments, like allowing gaps in sections of fencing, such that wildlife does not get trapped in residential yards or similar sites and can easily find their way through neighborhoods and public spaces. In other cases, larger-scale projects of human design can more explicitly serve as corridors, such as highway underpasses and overpasses and greenways (Figures 2.2, 2.3, and 2.4).

When considering functional connectivity from the perspective of individual species, it is also important to remember that what enables functional connectivity for one species might not work as well for another. Here, Hilty et al. note the example of a cliff; that is, a cliff may be "perfectly manageable for a lizard to move up the rocks, but impassable to (most) humans."[33] With this, it can also be challenging to gauge functional connectivity because "it is species-specific and depends not only on the species' movement ability

KEY CONCEPTS AND IDEAS 19

Figure 2.2 A breach in corral-style fencing, likely created by deer passing through this residential area, was intentionally left as is by the landowner to allow deer and other wildlife to more easily exit the property. Credit: Michael Propen.

Figure 2.3 Photograph of a wildlife underpass crossing culvert for animals under a highway in the Netherlands. Credit: CreativeNature_nl.

Figure 2.4 Photograph of an overpass over the Trans-Canada highway designed for wildlife migration/crossing. Credit: Steve_Gadomski.

in relation to the landscape features but also on other factors such as the individual's internal motivation to move and the level of risk encountered when traveling."[34] Research methods for studying functional connectivity can acknowledge and mitigate these factors to some extent.

As mentioned earlier, structural and functional connectivity are interconnected and often inform each other in different ways; that is, for some bird species, "habitat patches may be functionally connected [...] without any visible landscape linkages," since birds are more able to move through a landscape matrix if the "expanse of matrix between habitat patches is not too vast."[35] In other agricultural contexts, structural and functional connectivity are more seamlessly intertwined; as Hilty et al. note, "when remnant forest patches in a matrix of fields and pastures are connected by hedgerows or riparian corridors, structural connectivity equates with functional connectivity for forest-dependent species."[36] Landscapes can also include certain elements that have the unintended outcome of functioning as corridors, especially when such landscapes have been left more undisturbed. Even when landscapes have been modified by human usage, they may still function as corridors to certain extents, although the outcomes are not always consistently positive. That is, some "vegetation structures within agricultural landscapes can function as movement corridors" for certain wildlife but simultaneously inhibit the movement of other species or inadvertently increase the presence of non-native species or predators.[37]

One example of unintended corridors that can have varied outcomes for species' movements are roadside corridors, or small strips of vegetation along roadsides. Such strips of vegetation along roadways can help provide connectivity for rodents and some birds, but they can also have the negative impact of attracting such species to areas of significant human use, where they become more susceptible to being struck by vehicles. More productively, undisturbed vegetation strips near embankments and streams "can serve as both a habitat and a conduit for species traveling among larger habitat patches."[38] It's therefore necessary, with any connectivity project, to work to avoid unintended negative consequences by evaluating how a corridor may impact different species with different needs (Figure 2.5).

Greenways, or open-space systems or greenbelts, can also function as corridors for wildlife. Greenways are typically human-designed spaces that serve multiuse purposes and are created with human recreation and wildlife and ecosystem protection in mind.[39] Boston's Emerald Necklace, designed by Frederick Law Olmsted in the 1900s, is perhaps one of the most well-known greenbelt systems within the United States. The Emerald Necklace Park System is one of the main land-based wildlife corridors in Boston today.[40] When Olmsted created the parks that comprise the Emerald Necklace system, he designed them to be "comprised of large expanses of woodland, meadow and water in which wildlife could thrive and be protected from intense human intervention"; in doing so, he designed the parks to stand out as much as possible from the surrounding urban environment[41] (Figure 2.6).

Figure 2.5 Aerial view of roadside vegetation. Credit: Wirestock.

Figure 2.6 Photograph overlooking Boston Common, Charles River, and Back Bay: part of the Emerald Necklace System. Credit: Christopher Marino.

Corridors as Green Infrastructure

Human-designed connectivity projects like greenways, overpasses, and underpasses are all forms of green infrastructure. In recent years, an increased movement toward green infrastructure projects arguably marks a paradigm shift in which humans are recognizing with greater frequency the need to coexist with vulnerable species. As the climate crisis gives rise to more frequent and severe storms and wildfires, species like mountain lions and black bears are venturing closer to areas of significant human habitation, largely because their home ranges are being destroyed by natural disasters. As a result, humans must account for more frequent encounters with wildlife in residential neighborhoods, tourist areas, and along roadways and interstates. Green infrastructure corridor projects that are designed with increased connectivity in mind are one way to mitigate these issues while drawing upon compassionate conservation and entangled empathy.

Green Infrastructure and the Florida Panther: Compassionate Conservation at Work

State Road 80 in LaBelle, Florida, is one of the deadliest roads in Florida for wildlife—especially for the Florida panther. Moreover, State Road 80 is about one mile away from the Caloosahatchee River, which is an important ecosystem for Florida panthers. To help prevent panther deaths and automobile accidents, as well as to reconnect Florida panthers with a critical portion of their ecosystem, the Florida Department of Transportation created an underpass that runs beneath a section of the State 80 highway and allows panthers to safely cross beneath the highway. Because of the presence of this underpass, Florida panthers in this area have been able to access the north side of the Caloosahatchee River for the first time in five decades.[42]

Florida panthers once thrived in the Southeastern United States, from the Carolina mountains to Louisiana, and throughout the Everglades. Like many charismatic megafauna, the Florida panther's numbers began to dwindle as human development encroached on their habitat.[43] By the late 1960s, there were fewer than 30 panthers left in the region, and they resided mostly around Okaloacoochee Slough in Southwest Florida, where they were totally isolated from their "more expansive territory to the north"; moreover, this dwindling population was plagued by genetic mutations like "kinked tails, heart-related disorders and fertility problems."[44] Nonetheless, human development continued, and residential neighborhoods and interstate systems continued to threaten their population; vehicle strikes in particular were taking a toll on these creatures.

In the mid-1990s, The Nature Conservancy (TNC) started a wildlife corridor project with the goal of protecting the Florida panther. The corridor would be comprised of connected conservation areas "that would allow panthers to escape populated areas in the southwest, cross the Caloosahatchee River (which runs from Lake Okeechobee to Fort Myers) and establish home ranges in the wide open areas of Central and North Florida."[45] Other stakeholders, like ranchers, government agencies, and conservation groups, came together to support the two main goals of increasing the panther population and expanding their habitat. By the late 1990s, TNC and its partners protected the "29,495-acre slough" and also secured additional critical panther habitat that included "the Big Cypress National Preserve, Everglades National Park and the Florida Panther National Wildlife Refuge near Naples."[46]

The Caloosahatchee River specifically was a key area to protect, for north of this area, ranches and other protected regions offered ideal habitat for these panthers. No female panthers had been seen beyond the river since the 1970s, and conservationists agreed that both sides of the river needed to be protected in order for the panthers to be able to cross it safely.[47] Between 2012 and 2015, The Nature Conservancy worked with state and federal agencies to protect the area and ultimately "acquired a conservation easement to protect more than 1,527 acres at Black Boar Ranch, with the assistance of the U.S. Department of Agriculture."[48] The Florida Department of Transportation also installed fencing along the busiest roads, as well as the underpass at

State Road 80.[49] Most recently, a camera trap recorded two kittens with their mother, which was the first such sighting in over 40 years; while this is a small step toward the establishment of a subpopulation, much work is left to be done in order to complete the wildlife corridor and safely remove the Florida panther from the endangered species list.

The ongoing story of the Florida panther not only helps illustrate a successful example of functional connectivity but also reflects the challenges of land management and the broad range of stakeholders involved in efforts to enhance biodiversity conservation. This regional wildlife corridor also has broader implications for compassionate conservation and understanding the impacts of human development on vulnerable species. Of course, the Florida wildlife corridor would not be possible without the cooperation and participation of many stakeholders. The importance of inclusivity in connectivity projects cannot be understated, and it has become even more recognized in recent years that communication inclusive of many organizations and groups is necessary for connectivity and corridor projects to work. For another positive example that also highlights the successful interplay of inclusivity and connectivity in corridor management, we may look to the Yellowstone to Yukon (Y2Y) Corridor and the conservation organization that helps advocate for and manage the region: the Y2Y Conservation Initiative.

The Yellowstone to Yukon (Y2Y) Conservation Initiative: "A Geography of Hope"

The Y2Y region encompasses approximately 3,400 kilometers (2,100 miles) from the Greater Yellowstone Ecosystem to Canada's Yukon Territory. It spans "five American states, two Canadian provinces, two Canadian territories, and at least 75 Indigenous territories."[50] The Y2Y Conservation Initiative is an important example of the ways in which traditional ecological knowledge helps shape the success of connectivity projects.

In the 1990s, researchers began to understand that wildlife required larger, more connected expanses of territory than previously thought in order to thrive. These shifting views informed approaches to understanding the Y2Y territory, and conservationists soon agreed that to ensuring the sustained connection and protecting larger expanses of habitat in this region would help give wildlife the room they needed to move through the area and prosper.[51] Not unlike the Florida wildlife corridor, in addition to protecting undeveloped habitats, the Y2Y Conservation Initiative focused on improving road safety, both for people and wildlife.[52] The Y2Y region soon became understood by stakeholders as "a geography of hope"—one that could ideally "support people, all wildlife and natural systems."[53] As such, it arguably models what Kimmerer would refer to as an "Indigenous worldview" about conservation and restoration—one that understands the ecosystem and the species that inhabit it as "a community of sovereign beings."[54] Significantly, the Y2Y Conservation Initiative has demonstrated the importance of including the voices of Indigenous peoples and traditional ecological knowledge in the management of conservation areas.

Early on, as the Y2Y Conservation Initiative's president and chief scientist Dr. Jodi Hilty describes, the organization focused primarily on land use and on identifying habitat most critical for species' migration; however, as the group soon discovered, "while conservation science could pinpoint the places that wildlife needed, protecting those places required forming relationships and building trust with and among the human communities of the Rockies— and that, in turn, required many years of meeting with and listening to local people."[55] Over two decades later, the support and participation of Indigenous peoples have led to new conservation agreements and an 80 percent increase in protected areas within the corridor, including regions that span "five American states, two Canadian provinces, two Canadian territories and at least 75 Indigenous territories."[56] Among these newer, Indigenous-led conservation agreements is a partnership with West Moberly First Nations and Saulteau First Nations (collectively, the "Nations") to protect caribou habitat within the Peace River region of British Columbia.

Caribou have long been critical to the subsistence and cultural values of Indigenous peoples across Canada, but their numbers have been steadily declining in recent years. In response to the declining population of mountain caribou in their home territory, the Nations "came together to create a new vision for caribou recovery on the lands they have long stewarded and shared."[57] In particular, the Nations focused on the "Klinse-Za subpopulation" of caribou; this subpopulation "declined from ~250 in the 1990s to only 38 in 2013," which made it impossible for Indigenous communities to harvest the caribou, thus "infringing on treaty rights to a subsistence livelihood."[58]

In collaboration with many groups and governments, including the Y2Y Conservation Initiative, this Indigenous-led conservation project combined an immediate population recovery plan with longer-term habitat protection planning with the hope of fostering a self-sustaining population of caribou. As a result of these efforts, the "Klinse-Za caribou has more than doubled from 38 animals in 2013 to 101 in 2021, representing rapid population growth in response to recovery actions."[59]

In 2020, after having successfully avoided the extinction of these caribou, the Nations then focused on "securing a landmark conservation agreement" to protect the caribou habitat "over a 7986-km^2 area."[60] This agreement ensures habitat protection for more than "85% of the Klinse-Za subpopulation" and also helps protect nearby caribou subpopulations; as a result, this "Indigenous-led conservation initiative" has enabled the Indigenous and Canadian governments to begin restoring the Klinse-Za subpopulation of caribou and, in the process, "reinstate a culturally meaningful caribou hunt."[61] Moreover, this conservation initiative demonstrates how Indigenous communities can implement "meaningful conservation actions, enhance endangered species recovery, and honor cultural connections to now imperiled wildlife."[62] As Hilty and Zenkewich further describe, "wildlife habitats cannot be reconnected in a meaningful and lasting way unless human communities of all kinds are also reconnected—linked to one another, as well as to their landscapes. By listening to and actively supporting community members and leaders, we can support all forms of life into the future, from Yellowstone to Yukon and beyond."[63] Given the increasing and unabating challenges of the Anthropocene, which include continued technological development that equates with human intervention into the landscape, as well as increasing natural disasters in a time of climate crisis, large-scale connectivity projects are more important than ever. Connectivity projects must be implemented in areas of significant human use, as well as in less-developed habitats that have been or will likely be impacted by the effects of climate change. These are areas of varying conservation status, such as areas of unlogged forest, regenerated areas, planted areas such as windbreaks or greenways, and along roadways and interstates. In each of these instances, the goals and approaches of connectivity projects must account for the needs of both people and wildlife, which means achieving a holistic picture of the various social, cultural, and economic contexts related to the habitats in question.

The next chapter will discuss in more depth the interdisciplinary challenges underpinning the conceptualization and management of connectivity projects, including the factors upon which landscape connectivity depends, such as understanding gaps in the habitat or the existence of other natural pathways that should be protected or leveraged. Moreover, as we have begun to discuss here, connectivity projects should account for and be informed by traditional ecological knowledge in their strategic vision.

Notes

1. Conservation Corridor, "Corridor FAQ."
2. Hilty et al., *Guidelines for Conserving Connectivity*, xii.
3. Hilty et al., *Guidelines for Conserving Connectivity*, xii.
4. Hilty et al., *Guidelines for Conserving Connectivity*, 14.
5. Hilty et al., *Guidelines for Conserving Connectivity*, 14.
6. Hilty et al., *Guidelines for Conserving Connectivity*, 15.
7. Hilty et al., *Guidelines for Conserving Connectivity*, 15.

8 Hilty et al., *Guidelines for Conserving Connectivity*, 15.
9 Conservation Corridor, "Corridor FAQ."
10 Anderson and Jenkins, *Applying Nature's Design*, 38.
11 Conservation Corridor, "Corridor FAQ."
12 Hilty et al., *Guidelines*, 40.
13 Hilty et al., *Guidelines*, 40.
14 Conservation Corridor, "Corridor FAQ."
15 Conservation Corridor, "Corridor FAQ."
16 Conservation Corridor, "Corridor FAQ."
17 Vermont Natural Resources Council, "Wildlife Corridor Protection."
18 Conservation Corridor, "Corridor FAQ."
19 Hilty et al., *Corridor Ecology*, 103.
20 Lewis Creek Association, "Monkton Wildlife Crossing."
21 National Park Foundation, "Saint Croix."
22 Two Countries, One Forest, 2019.
23 Anderson and Jenkins, *Applying Nature's Design*, 23.
24 Anderson and Jenkins, *Applying Nature's Design*, 23.
25 Anderson and Jenkins, *Applying Nature's Design*, 23.
26 Meyer et al., "Towards the Restoration of the Mesoamerican Biological Corridor for Large Mammals in Panama," 1.
27 Meyer et al., "Towards the Restoration of the Mesoamerican Biological Corridor for Large Mammals in Panama," 1.
28 Meyer et al., "Towards the Restoration of the Mesoamerican Biological Corridor for Large Mammals in Panama," 3.
29 Meyer et al., "Towards the Restoration of the Mesoamerican Biological Corridor for Large Mammals in Panama," 3.
30 Meyer et al., "Towards the Restoration of the Mesoamerican Biological Corridor for Large Mammals in Panama," 2.
31 Meyer et al., "Towards the Restoration of the Mesoamerican Biological Corridor for Large Mammals in Panama," 2.
32 Hilty et al., *Corridor Ecology*, 94.
33 Hilty et al., *Corridor Ecology*, 94.
34 Hilty et al., *Corridor Ecology*, 94.
35 Hilty et al., *Corridor Ecology*, 94.
36 Hilty et al., *Corridor Ecology*, 94.
37 Hilty et al., *Corridor Ecology*, 98.
38 Hilty et al., *Corridor Ecology*, 95.
39 Hilty et al., *Corridor Ecology*, 98.
40 Terrascope, "Boston Background."
41 Pressley Associates, Inc., *The Emerald Necklace Parks*, 18.
42 The Nature Conservancy, "Filming the Ghost Cat."
43 The Nature Conservancy, "Florida Panthers."
44 The Nature Conservancy, "Florida Panthers."
45 The Nature Conservancy, "Florida Panthers."
46 The Nature Conservancy, "Florida Panthers."
47 Seeger, "The Panther's Path."
48 The Nature Conservancy, "Florida Panthers."
49 The Nature Conservancy, "Florida Panthers."

50 Y2Y Conservation Initiative, "Our Work."
51 Y2Y Conservation Initiative, "History."
52 Y2Y Conservation Initiative, "History".
53 Y2Y Conservation Initiative, "Our Work."
54 Kimmerer, *Braiding Sweetgrass*, 331.
55 Hilty and Zenkewich, "The Future."
56 Hilty and Zenkewich, "The Future."
57 Lamb et al., "Indigenous-led Conservation."
58 Lamb et al., "Indigenous-led Conservation."
59 Lamb et al., "Indigenous-led Conservation."
60 Lamb et al., "Indigenous-led Conservation."
61 Lamb et al., "Indigenous-led Conservation."
62 Lamb et al., "Indigenous-led Conservation."
63 Hilty and Zenkewich, "The Future."

Chapter 3

DESIGNING AND MANAGING WILDLIFE CORRIDORS

Just east of Salt Lake City, along Route I-80 in Parleys Canyon, Utah, is the recently completed Parleys Summit wildlife crossing. Completed in winter 2018, the crossing at Parleys Summit is the largest wildlife crossing in Utah. The interstate overpass is 50 feet wide and 320 feet long, spanning six lanes of I-80.[1] Three miles of fencing along both sides of the highway create a pathway that guides species to the overpass. Along the overpass, video technology and camera traps record the sights and sounds of "migrating moose, elk, deer, and other animals" as they traverse the freeway that has fragmented their local ecosystem. The overpass has also been used by some unexpected species, such as "bobcats, cougars, coyotes and a yellow-bellied marmot," and one Utah Department of Transportation (UDOT) spokesperson notes: "It's great to see so many different animals using the overpass."[2]

While the site is still too new to provide any long-term data, the species' initial responses to the crossing have been encouraging. Local media articles about Parleys Summit convey a range of perspectives on the project and are typically quick to note that the crossing was designed with the safety of both humans and animals in mind. The local organization Save People Save Wildlife had advocated for the crossing since 2016, largely out of concern that "the moose population was beginning to dwindle and drivers were at risk along I-80."[3] Similarly, a spokesperson for UDOT expresses: "It's exciting to have this done. [...] This has been the most talked about UDOT project of the year, rightfully so. It is unique and it is really going to improve the safety of drivers in Parleys Canyon by cutting down on the wildlife and vehicle collisions. I think it is really going to make a big difference."[4] While UDOT also acknowledges the benefits to wildlife, they ultimately emphasize the benefits to humans:

> We have a lot of wildlife in that area, including deer, elk and moose. [...] We obviously want to ensure their safety. But, the real purpose of this crossing is to ensure the safety of everyone traveling in the canyons.

They are the real beneficiaries. They are going to be able to drive in that area and not have to worry about wildlife coming onto the freeway.[5]

The Utah Division of Wildlife Resources has also been circulating videos that show animals using the overpass. In one tweet, the division says, "We're excited to see #wildlife using the new Parleys Summit overpass! The overpass and roadside fencing are providing safer migration routes and reducing collisions."[6] Thus, although there is clearly an expressed interest in ensuring the safety of wildlife, the primary motivation for the corridor seems to be driver safety.

My goal in describing the Parleys Summit crossing is to illuminate some emergent perspectives and questions that may apply more broadly to the planning, design, and stakeholder rationales for wildlife corridor projects. That is, the rationales for these projects often prompt us to consider our relationships with wildlife and the extent to which our rationales may be more focused on the needs and interests of humans or nonhuman animals. Further, if driver safety is part of the UDOT rationale for creating a wildlife overpass, do such human-centric motivations necessarily negate the benefits for wildlife? Does a human-centric rationale always need to be present in order to also protect wildlife? And, if it does, but if wildlife are also protected in the process and ideas about coexistence garner public support, are these multiple rationales necessarily a negative thing? As the forces of human development and climate change bring humans and wildlife in closer proximity to one another, we would do well to contemplate the implications of our decision-making and rationales about how to best coexist in vulnerable or fragmented ecosystems.

Rationales for Valuing Wildlife and Ecosystems

Wildlife corridor projects provide a unique window into the various ways that stakeholders perceive and value wildlife. These rationales then shape our ideas and communication practices about how various species may be impacted by our attempts to reconnect fragmented habitats. Communication scholar Julia Corbett, for example, considers and complicates the different types of attitudes that humans have regarding our relationships with nonhuman animals. Corbett cites *utilitarian* attitudes, involving the "practical and material value" of an animal's habitat; *naturalistic* attitudes, involving an "interest and affection for wildlife and the outdoors"; *ecologistic* attitudes, involving concern for the environment as a system and the "interrelationships between wildlife and habitats"; and *moralistic* attitudes, which would involve "concern for right/wrong treatment" and would "oppose cruelty and exploitation."[7]

It is worth noting that utilitarian approaches or perspectives can function in different ways, and to different ends. For example, a compassionate conservation approach would likely not endorse the U.S. Army Corps of Engineers' recent plan to kill 11,000 double-breasted cormorants in Oregon's Columbia River Estuary to protect local salmon populations.[8] Such an approach not only places value on one species over another but also enacts a "utilitarian" perspective that deems it acceptable to kill species viewed as "pests" due to their high populations within a given area.[9] Such vantage points run counter to the main principles of compassionate conservation. However, this practice is clearly different from suggesting that there seems to be a strong utilitarian component in the creation of the Parleys Summit crossing and other similar projects, in which case, the utilitarian function is more so that to prevent animals from entering the highway, thereby saving their lives, also helps prevent traffic accidents with vehicles, thereby saving human lives as well. In this case, the corridor project functions to protect the lives of both people and wildlife and helps reconnect fragmented ecosystems, thereby also improving biodiversity in the process.

These different belief systems and approaches, then, are complex and not mutually exclusive, and it may be possible to hold both utilitarian and ecologistic beliefs simultaneously. In particular, what Corbett describes as moralistic attitudes may be viewed as aligned with the compassionate conservation tenet of "do no harm" and our perceived ethical obligation to species. Likewise, Marc Bekoff, too, feels that "[w]e need to be kind and empathic and cooperate with one another, so that we can define and work toward common goals, even when we disagree on the exact path."[10] In the next chapter, I discuss further these emerging theoretical perspectives and how they may complement or inform ideas about the design and creation of wildlife corridors. While some critics have suggested that empathy and compassionate conservation are incompatible with the goals of conservation biology, I suggest there may be a middle ground in how we understand these approaches, as it pertains to the work of wildlife corridors.[11] In fact, many of the stakeholders who speak on behalf of these projects implicitly articulate a combination of values related to the need for wildlife corridors.

The conceptualization, design, and management of connectivity projects are thus complex and informed by many factors, and the goals or outcomes of their design—the ultimate benefits to stakeholders—depend on how these factors are understood. The more overtly instrumental goal of reducing traffic accidents, for instance, applies primarily to connectivity projects focused on interstate underpasses and overpasses, but many connectivity projects exist in areas where building highway overpasses would not provide a relevant solution. In such cases, other factors affecting landscape connectivity also apply, such as the need to protect or leverage already existing natural pathways. Wildlife corridors may, when possible, attempt to account for the best interests of multiple stakeholders, which often encompass both people and wildlife; however, it is often easier said than done to meet the needs of multiple stakeholders who may also have different or even competing interests. With these ideas in mind, this chapter considers the contexts not only for understanding the implementation of wildlife corridors but also for the interdisciplinary challenges underpinning the ongoing management of corridor projects. It provides some general design guidelines while also keeping in mind that every project is unique in its context; nonetheless, corridor design should strive to leverage existing, natural corridors, mitigate climate change when possible, and minimize the connecting of disturbed habitats with less-disturbed habitats.

Connectivity Challenges: Understanding Obstacles and Stakeholder Needs

Corridor design is always context-specific and depends on multiple factors, including species, habitats, and the surrounding ecosystems. Design not only needs to account for the "biophysical elements of a corridor but the socioeconomic and political factors influencing corridor configuration and implementation."[12] There are several cited obstacles to corridor implementation and management, which include lack of awareness or understanding, perceived lack of control over resources, cost, and uncertainty.[13] To combat lack of awareness and understanding specifically, Anderson and Jenkins advocate for engaging as wide a range of stakeholders as possible, and for educating audiences about landscape connectivity and conservation issues. They note that it is often difficult to conceptualize the problems that corridors address, such as habitat fragmentation and biodiversity loss, "especially at large geographic scales where broad public support for corridors is most critical."[14] To broaden public support, Anderson and Jenkins recommend having knowledge of "both potential allies and adversaries of the initiative, which can be achieved through a stakeholder analysis."[15] Such an analysis is critical in any project where communication is key to forward movement. In this case, a stakeholder analysis should identify the key players or interest groups and their knowledge and values, especially those related to environmental issues and the corridor specifically; how those groups may feasibly contribute to the corridor project or even undermine it; and strategies for gaining their approval or lessening their resistance.[16] Such issues of audience analysis and persuasion show that the design, implementation, and management of wildlife corridors have a clear social and political component in addition to understanding the conservation needs of species and ecosystems; thus, clear communication across audiences is key to gaining support for corridor projects. In order to address any potential opposition early on, it is also necessary to consider whether a corridor is the best solution in the first place; for instance, the creation of a more "narrow, linear corridor" might not be the best solution when it is possible to first increase "the area of existing reserves."[17] In this way, an approach that involves less disturbance of the existing landscape would be preferable to one that alters the landscape.

That is, as Anderson and Jenkins describe, it's often preferable "to preserve isolated sites of high conservation value than to acquire more accessible but marginal habitat for corridors."[18] Moreover, the condition and uses of the landscape are also large factors affecting design. That is, if a landscape is already highly vulnerable, then it might be safer to link existing fragments with smaller-scale, linear corridors. When a landscape is intact, then the focus might be on "protecting the predominant habitat matrix."[19] With regard to the Y2Y region described in the previous chapter, conservation work in that region of the southern Rocky Mountains looked more at maintaining or reviving connectivity in an already-developed landscape. Moreover, the Y2Y corridor project achieved this connectivity in a way that not only preserved native landscapes but also allowed human communities to thrive—as such, it is "inclusive of all beings, human and nonhuman."[20]

Connectivity Goals: Account for the Needs of Various Stakeholders as Much as Possible, Including Wildlife

No matter the context and landscape, it is necessary to have an understanding of the larger goals of a corridor project. Moreover, in order to meet the needs of various stakeholders and to best leverage available resources, corridor projects should ideally have multiple goals in mind and serve purposes across audiences if possible. That said, a corridor's multiple uses should not contradict or counteract one another, which makes it all the more important that the stakeholders who are managing the project communicate with each other to understand all of its contexts and goals.

The Parleys Summit wildlife crossing explicitly serves the needs of both people and wildlife and leverages existing landscape features. In Florida, the human-made underpass that runs beneath the section of the State 80 highway, referred to as the Caloosahatchee Ecoscape and discussed in the previous chapter, is integrated into the existing landscape but connects two fragmented patches, thus serving as a wildlife corridor that helps prevent automobile accidents as well as reconnect Florida panthers with a critical portion of their ecosystem and prevent panther deaths in the process.

Florida panthers are a good example of what is called a "flagship species," or "a charismatic animal that helps attract public support for conservation."[21] Traditionally, wildlife corridors tend to be designed based on the needs of such flagship or keystone species. Along similar lines, a "keystone species" is a species that often serves as a barometer of an ecosystem's health. The California sea otter, for example, is not only considered a charismatic, flagship species but also a keystone species that provides a window into ocean health in areas of coastal California; this is because the sea otter is a "top predator of invertebrates along the California coast," and so any "changes in the health of its population can make scientists aware of variations in the ocean environment itself."[22]

Conservation science has traditionally used the requirements of keystone and flagship species as a starting point when determining the size and design of a wildlife corridor. Such an approach relies on the logic that if a corridor meets the needs and connectivity goals of the largest predators, it will, in turn, be likely to meet the needs of additional species.[23] From a societal perspective, this approach can also help garner public support for wildlife corridor projects, much as it did for the Florida panther and the Caloosahatchee Ecoscape. In order to best engage in compassionate conservation, however, corridor design should account not only for flagship and keystone species but also, ideally, for the needs of as many species in the habitat as possible, because, as described earlier, what serves as a corridor for one species may not necessarily work as well for another. For, in one study, as Hilty et al. have described, the "overlap in dispersal habitat of a bird, a butterfly, and a frog in a fragmented landscape was high for places identified as important for dispersal, suggesting a corridor designed for one of these taxa will serve the other two as well."[24] On the other hand, a wildlife corridor plan in California that was designed around the "life history" of mountain lions was found to support several valuable biodiversity elements, such as "serpentine rock as a surrogate for rare plants, old-growth forest, different types of oak woodlands, and watersheds containing endangered fish," but at the same time, researchers discovered that "endemic amphibian, reptile, and mammal populations were not well represented by the path of the puma."[25] The relatively "poor umbrella function of carnivores for corridor design" has also been noted in another study that compared "corridors between protected areas," and so using the needs of umbrella species as a starting point for corridor design should be "carefully considered."[26] That said, to garner public support for vulnerable species is also one method of engaging an entangled empathy approach, in which members of the public have an opportunity to learn more about the landscape from the perspective of wildlife and vulnerable species. And, as is the case with the Florida panther, as discussed in Chapter 2, sometimes a corridor project is exclusively focused on the needs of a particular species, especially when that species is on the endangered species list or completely cut off from critical habitat. Even so, to plan and design a wildlife corridor also arguably requires knowledge of all impacted species apart from solely umbrella species; this depth of knowledge can then arguably help foster greater empathy for a wide range of species.

Connectivity Goals: Minimize Disturbances to Landscapes as Much as Possible, and Balance Biodiversity Needs with Other Contexts

Prior to creating new corridors, then, the existing ones should be evaluated and leveraged when possible. This helps minimize habitat disturbances and reduce stress on species and landscapes by reducing any unnecessary human interventions. In some cases, however, it is not always possible to completely avoid disturbances to species in habitat reconstruction projects; the issue of biodiversity offsetting reflects one such challenge.

Balancing Biodiversity Needs with Other Contexts: The Challenges of Biodiversity Offsetting

The need to balance biodiversity issues with social and cultural contexts is complex, as is illustrated by biodiversity offsetting, broadly defined, biodiversity offsets "are measurable conservation outcomes designed to compensate for adverse and unavoidable impacts of projects, in addition to prevention and mitigation measures already implemented."[27] More specifically, biodiversity offsetting involves "the procedure of compensating for the residual loss or harm caused to nature by human activity by taking restoration or conservation actions in another location."[28] Put more simply, biodiversity offsetting can involve compensating for habitat losses in one area, which may result from a construction or development project, by creating or fostering new, similar habitats in a nearby area. For instance, the Chiltern Railways case discussed later in this book describes an offsetting example in which some upgrades to railway tracks also involved the construction of "purpose-built habitats" alongside the railway.[29] These habitats were designed for use by the great crested newts "that were using ditches, ponds and terrestrial habitat on and adjacent to the [existing] railway embankment."[30] In order for the railway upgrades to proceed, the railway company had to ensure that the newts "would be unaffected by the development, and that reasonable measures would be set in place to avoid killing or injury" of them.[31] The subsequent mitigation strategy then entailed "the creation and enhancement of ponds and the construction of hibernacula and refugia to replace those to be removed or damaged."[32] The newts were then moved from the development area and translocated to the previously created receptor areas.[33] As the ecology consulting firm involved with the project describes, this undertaking required consistent and detailed communication with all stakeholders involved: "Throughout the life of the project, the complexities of ownership and responsibilities in connection with the project required extensive liaison with Natural England, tenants, landowners, clients and other stakeholders. Allocating an experienced ecologist dedicated specifically to the project has been essential in maintaining consistency and timeliness of sound advice to stakeholders."[34] Biodiversity offsetting is one area of conservation biology that can challenge some of the principles of compassionate conservation, depending on the project, scale, and level of habitat loss involved; however, this is not necessarily meant to suggest that such measures be ruled out entirely when it comes to corridor design and implementation. Rather, it means thinking critically about what land means to us. As Kimmerer points out:

> How we approach restoration of land depends, of course, on what we believe "land" means. If land is just real estate, then restoration looks very different than if land is the source of a subsistence economy and a spiritual home. Restoring land for production of natural resources is not the same as renewal of land as cultural identity. We have to think about what land means.[35]

Likewise, this book advocates for awareness, understanding, compassion, nuance, and accounting for the voices of all stakeholders—especially at the most local levels—when it comes to the management of fragmented habitats and corridor projects. It also advocates for thinking critically about what land means and for whom.

Connectivity Goals: Engage in a Compassionate Conservation Approach

To incorporate the principles of compassionate conservation into corridor design also means understanding the behaviors, patterns, and requirements of the specific species who would use the corridor. That is, the more that is known about a species, the greater the likelihood that they will "find, select, and successfully pass through a corridor."[36] If wildlife are unable to travel the required distance to locate the corridor, they will not be able to use it; if they do not perceive the corridor habitat to be of adequate quality or preferable to the fragmented, patch matrix, then they will be less likely to use it; if they are unable to make it safely to the corridor without encountering a predator, they will not be able to safely use the corridor.[37] All of these factors rely on understanding the needs and behaviors of the species in question, and such understandings are related to an empathetic, compassionate conversation approach.

Knowledge of species' social organization as well as their dispersal ecology is an important factor in determining corridor design. With regard to dispersal ecology, "many animal species may be most likely to utilize corridors during dispersal, when young animals leave the maternal or natal home-range to establish their own territories."[38] Dispersal ecology entails understanding the "timing, direction, and distance" of a species' migration and is thus relevant for successful corridor design (31). Some species that are at high risk for predation, like White-tailed deer, will disperse quickly "through short, rapid transfers between groups."[39] Larger carnivores, like mountain lions, will disperse farther when they have the opportunity to do so. Black bears are highly reliant on their social group and will tend to stay within an overlapping distance of their maternal home-range.[40]

Connectivity Goals: Climate Considerations

The increasingly dire climate crisis also factors into corridor design considerations now more than ever before. As landscapes change in response to increasing natural disasters, species' migration patterns will in turn be impacted. While corridors may help mitigate such problems, this puts pressure on wildlife corridors to be of sufficient quality such that they can withstand climate impacts; thus, "conservation of large tracts of high-quality habitat in landscape corridors provides the best hedge against climate change impacts."[41] Designing movement corridors in response to current or anticipated climate change is challenging due to the predictive uncertainties involved.

When designing movement corridors in response to climate change, Anderson and Jenkins also note that corridors should create a wide network that affords migration in multiple directions in order to provide greater amounts of habitat and larger movement pathways for species with "relatively low mobility, such as many plants and terrestrial invertebrates."[42] Again, as the recent 2020 IUCN report also notes: "Science overwhelmingly shows that interconnected protected areas and other areas for biological diversity conservation are much more effective than disconnected areas in human-dominated systems, especially in the face of climate change."[43] Here, management strategies that involve ongoing monitoring become even more necessary to help ensure that corridors continue to align with environmental changes wrought by climate change over time.

Corridor Design Elements and Considerations

Traditionally, ideas and recommendations related to corridor design have focused on the three interrelated structural elements of width, connectivity, and habitat quality.[44] As it pertains to corridors, width is related to "how much of the corridor interior is exposed to disturbances or edge effects, whether natural or human-induced, from the surrounding matrix. Connectivity refers to the degree to which gaps interrupt corridor habitat. Quality depends on both width and connectivity and reflects how closely the corridor approximates pristine habitat."[45] The effectiveness of these elements is then impacted by landscape features such as river and stream systems or riparian corridors, which vary greatly depending on ecosystems and other climate and usage factors.

Broadly speaking, there is general agreement within the conservation science community that "corridors designed for large, highly mobile species vulnerable to human disturbance should be as wide as possible."[46] Habitat quality also influences width, and ensuring habitat of high enough quality means providing enough width: "to protect interior habitat from edge effects and to accommodate small-scale natural disturbance and succession."[47] Gauging the relative success and appropriateness of width and quality requires knowledge and observation of the species that must traverse the corridor in question. Likewise, the connectivity required of a corridor is highly dependent on knowledge of the species who will use it and the amount of gap versus contiguous habitat that those species find tolerable.

When corridors can help connect diverse landscapes, there is generally an increased likelihood of fostering greater biodiversity within that corridor. Different species will make use of landscape features in different ways, and so it follows that the more landscape features that can be connected within a corridor, the greater the likelihood that various species will be able to make use of that corridor. River and stream systems in particular can invite, encourage, and concentrate biodiversity by serving as conduits for species and other terrestrial matter.[48] Riparian corridors, or communities of plants and vegetation that grow nearby rivers, streams, or other natural bodies of water, can help "preserve watersheds and aquatic ecosystems that benefit human communities."[49] Riparian corridors are crucial for both people and wildlife, and serve important functions such as "[p]reserving water quality by filtering sediment from runoff before it enters rivers and streams, [p]rotecting stream banks from erosion, [p]roviding a storage area for flood waters, [p]roviding food and habitat for fish and wildlife, [and] [p]reserving open space and aesthetic surroundings."[50] To maximize biodiversity, a riparian corridor should provide areas of natural vegetation wide enough to accommodate "the geomorphic floodplain, and the headwater and groundwater sources that feed the waterway."[51] Slopes, croplands, and deforested areas are all landscapes in which riparian vegetation may thrive.

As Anderson and Jenkins describe, there is general consensus within the conservation science community that attention to the following five general design guidelines "can increase the biological value of corridors and help minimize the potential negative impacts" to landscape and species. For more detailed discussion, please see Anderson and Jenkins 2006, Hilty et al. 2019, and Smith and Hellmund, 1993:

Five General Design Guidelines for Corridors

- Focus on connecting patches that were previously connected and contain "naturally contiguous habitat type."
- Avoid connecting artificial patches with less-disturbed, higher-quality habitat.
- Seek out and preserve currently existing "natural corridors such as riparian zones and migration routes," for riparian zones in particular can help preserve water quality and maintain biodiversity, especially in arid regions.
- Situate corridors on altitudinal and latitudinal gradients to achieve the greatest possible biodiversity and help mitigate climate impacts.
- Avoid long expanses of landscape (greater than one mile or two kilometers) without nodes or endpoints like habitat patches, and incorporate redundant connections through the use of alternate pathways.[52]

The broad guidelines noted above should be considered within the specific context of the ecosystem in question and the conservation goals of the individual corridor project. Such goals should, again, be as specific as possible; that is, what specific species or groups of species should the corridor account for, and in what way? Should the corridor provide new habitat, or facilitate movements like migration or dispersal? Once these and related questions are sorted out, the "candidate area" may be more specifically determined, and related issues such as stakeholder concerns may be considered. Finally, to these design guidelines, I would also add that critical awareness of what land means and for whom should be considered in the specific context. For, as Kimmerer reminds us, "like other mindful practices, ecological restoration can be viewed as an act of reciprocity in which humans exercise their caregiving responsibility for the ecosystems that sustain them."[53] In this way, traditional ecological knowledge (TEK) also becomes an integral part of corridor planning and design.

Balancing Multiple Perspectives and Sources of Knowledge

The sociocultural and political issues affecting corridor planning and design require communication across groups and audiences in order to make sure that all parties are on the same page and that priorities for conservation goals are aligned and represented as much as possible. That is, a smaller, linear corridor may involve discussions of property development rights such as easements or the adherence to zoning codes. Of course, delicate issues related to land rights should be approached with care and open, inclusive communication, as the Y2Y project has demonstrated. Likewise, larger, more regional projects may entail the need for larger uptake or public support and the need for a broader participatory approach. In either case, the need to be attentive to issues of land rights and local knowledge is paramount. In addition, unique management scenarios may also arise in marine environments "because issues there are often different than on land, where rights may be relatively clear. [...] In many countries, coastal communities may own or have tenure use rights over certain marine areas or resources."[54] The Papahānaumokuākea Marine National Monument (PMNM) case discussed later in this book helps illustrate how complex governance arrangements in marine corridor areas can be carried out successfully.

While unclear communication can lead to poor planning or even conflict, clear communication, guidelines, and expectations can have more positive outcomes, as the case involving Hawaii's PMNM shows. And, as the Y2Y conservation project and the PMNM cases in particular demonstrate, the need to be attentive to and incorporate TEK is key to the success of connectivity projects. The Y2Y project has also demonstrated that it is possible to balance conservation needs and ecological value systems with compassion and empathy for both people and wildlife. The next chapter will discuss in more detail some of these emerging theoretical perspectives and how they can both challenge and align with wildlife corridor projects.

Notes

1. McNaughton, "UDOT."
2. McNaughton, "UDOT."
3. Pierce, "New $5 Million, Animals-Only Overpass."
4. McNaughton, "UDOT."
5. McNaughton, "UDOT."
6. @UtahDWR, "We're Excited!"
7. Corbett, *Communicating Nature*, 190.
8. Anderson, "Audubon."
9. Tobias, "Compassionate Conservation."
10. Bekoff, *Animal Manifesto*, 199.

11 See Griffin et al., "Compassionate Conservation Clashes with Conservation Biology."
12 Anderson and Jenkins, *Applying Nature's Design*, 27.
13 Anderson and Jenkins, *Applying Nature's Design*, 55.
14 Anderson and Jenkins, *Applying Nature's Design*, 56.
15 Anderson and Jenkins, *Applying Nature's Design*, 63.
16 Anderson and Jenkins, *Applying Nature's Design*, 63.
17 Anderson and Jenkins, *Applying Nature's Design*, 27.
18 Anderson and Jenkins, *Applying Nature's Design*, 27.
19 Anderson and Jenkins, *Applying Nature's Design*, 28.
20 Bekoff, *Rewilding*, 9.
21 Anderson and Jenkins, *Applying Nature's Design*, 30.
22 Monterey Bay Aquarium, "Sea Otters."
23 Anderson and Jenkins, *Applying Nature's Design*, 30.
24 Hilty et al., *Corridor Ecology*, 118.
25 Hilty et al., *Corridor Ecology*, 118.
26 Hilty et al., *Corridor Ecology*, 120.
27 IUCN, "Biodiversity Offsets."
28 Tupala et al., "Social Impacts of Biodiversity Offsetting: A Review," 1.
29 Railway Technology, "Railways and Wildlife."
30 BSG Ecology, "East West Rail Ecology Mitigation."
31 BSG Ecology, "East West Rail Ecology Mitigation."
32 BSG Ecology, "East West Rail Ecology Mitigation."
33 BSG Ecology, "East West Rail Ecology Mitigation."
34 BSG Ecology, "East West Rail Ecology Mitigation."
35 Kimmerer, *Braiding Sweetgrass*, 328.
36 Anderson and Jenkins, *Applying Nature's Design*, 31.
37 Anderson and Jenkins, *Applying Nature's Design*, 31.
38 Anderson and Jenkins, *Applying Nature's Design*, 31.
39 Anderson and Jenkins, *Applying Nature's Design*, 32.
40 Anderson and Jenkins, *Applying Nature's Design*, 32.
41 Anderson and Jenkins, *Applying Nature's Design*, 35.
42 Anderson and Jenkins, *Applying Nature's Design*, 35.
43 Hilty et al., *Guidelines*, xii.
44 Anderson and Jenkins, *Applying Nature's Design*, 35.
45 Anderson and Jenkins, *Applying Nature's Design*, 36.
46 Anderson and Jenkins, *Applying Nature's Design*, 37.
47 Anderson and Jenkins, *Applying Nature's Design*, 38.
48 Anderson and Jenkins, *Applying Nature's Design*, 41.
49 Anderson and Jenkins, *Applying Nature's Design*, 41.
50 County of Santa Cruz Planning Department, "What Is."
51 Anderson and Jenkins, *Applying Nature's Design*, 42.
52 Anderson and Jenkins, *Applying Nature's Design*, 43.
53 *Braiding Sweetgrass*, 336.
54 Hilty et al., *Guidelines*, 31.

Chapter 4

EMERGING THEORETICAL PERSPECTIVES

Compassionate Conservation, Empathy, and Traditional Ecological Knowledge

This chapter explores some of the emerging theoretical perspectives that I suggest should inform and complement wildlife corridor projects. In it, I describe the ways that wildlife corridor projects may benefit from the inclusion of perspectives grounded in compassionate conservation, entangled empathy, and traditional ecological knowledge (TEK). Subsequently, this chapter argues for an approach to connectivity and coexistence that rethinks more commonly perceived boundaries and hierarchies in human/nonhuman animal relationships, that works against anthropocentric and hierarchical thinking, and that does not necessarily privilege humans above other species, or privilege one kind of species over another. In short, we must design spaces that allow for people and wildlife to coexist in the Anthropocene.

Compassionate Conservation

Compassionate conservation is an interdisciplinary movement that has steadily gained international attention and momentum in recent years. A compassionate conservation approach, as the University of Technology Sydney's Centre for Compassionate Conservation describes it, "promotes the treatment of all wildlife with respect, justice, and compassion" and is based on the guiding principles of *"first, do no harm, individuals matter, inclusivity, and peaceful coexistence.*" Moreover, it "aims to find solutions for conservation practitioners that minimise harming wildlife."[1] When applied to the practice of wildlife corridors, compassionate conservation would advocate for an approach that takes into consideration the lives of all species that may utilize or migrate through specific habitats. It would recognize that wildlife may be affected by the actions of humans, whether intentional or unintentional, as well as by the natural processes affecting the ecosystems that wildlife inhabit. It would then seek to minimize any harm to wildlife to the extent possible, regardless of the purpose behind the action.

Marc Bekoff, a scholar of ecology, evolutionary biology, and animal behavior who initially conceptualized compassionate conservation, likewise advocates for such an approach. He reiterates that the initial step in enacting compassionate conservation is to do no harm, but he also emphasizes a necessary shift in perspective that recognizes animals as sentient, individual beings:

> It's critical to avow that sentience matters. Science tells us animals have feelings, emotions, and preferences and individuals care about and worry about what happens to them and to their families and friends. We need to consider what we know about animal sentience when we intrude into their lives, even if it is on their behalf. [...] A humane framework that considers individual animals is long overdue.[2]

The "humane framework that considers individual animals" that Bekoff describes ought not to be mistaken as an interest in trading humanism for anti-humanism. Rather, compassionate conservation may be understood as a philosophy of accountability that calls into question the hegemonic power structures that underpin human exceptionalism or places rigid, hierarchical frameworks on how we value the lives of nonhuman animals. For, as discussed in the previous chapter, designing wildlife corridors solely around the needs of umbrella species does not always lead to the most optimal outcomes. Bekoff also notes that the differences in how we understand the rights and lives of animals are enormous and can lead "to very different priorities about who lives, who dies, and why."[3] Bekoff makes the following point, very directly put, when he advocates for a more nuanced and critically aware approach to wildlife management:

> Whenever humans seek to "manage" nature, creating parks and artificial boundaries, it is always only for the benefit of humans. Perhaps, to the degree to which animals are left alone within these parks, it might be said that animals benefit, that they have been protected from humans. Otherwise, most of what passes for "wildlife management" looks like nothing so much as a direct attack on wildlife itself.[4]

Instead, Bekoff advocates for what he calls "rewilding projects," or conservation projects that eschew a perceived hierarchy of species or that place borders between humans, animals, and the natural world.

Rewilding

As Bekoff describes, rewilding is not about the domination of humans over nature, or even of the sciences over the social sciences or humanities: "It reflects the desire to (re)connect intimately with all animals and landscapes in ways that dissolve borders. Rewilding means appreciating, respecting, and accepting other beings and landscapes for who or what they are, not for who or what we want them to be."[5] Consistent with the understanding that designing wildlife corridors solely around the needs of umbrella species, with certain exceptions, does not always lead to the most optimal outcomes, Bekoff's rewilding strategy, when applied to corridors and conservation, is based on the three premises that "(1) healthy ecosystems need large carnivores, (2) large carnivores need big, wild roadless areas, and (3) most roadless areas are small and thus need to be linked."[6] He invites conservation biologists to conceptualize rewilding as "a large-scale process involving projects of different sizes that may focus on carnivores but ultimately include a panoply of wildlife."[7] The Yellowstone to Yukon (Y2Y) Conservation Initiative and, I would add, the U.S. Highway 93 North reconstruction project that runs through the Flathead Indian Reservation in Montana, discussed later in this book, meet these criteria and provide examples of how humans can coexist with wildlife, with a mindset grounded in compassion and empathy.

Entangled Empathy

Rewilding and a compassionate conservation ethic are not incompatible with ideas about what philosopher Lori Gruen refers to as "entangled empathy." Entangled empathy, or "being able to understand what another being feels, sees, and thinks, and to understand what they might need or desire," Gruen says, "requires a fairly complex set of cognitive skills and emotional attunement."[8] Because entangled empathy focuses on "another's experiential wellbeing," this kind of empathy tends to lead to action based on an assessment of what seems to be the best, or most compassionate choice, in helping to pursue the wellbeing of another.[9] Gruen's theory of entangled empathy is grounded in an understanding of the entangled relationships of humans with other nonhuman animals. She argues that these relationships "co-constitute who we are and how we configure our identities and agency, even our thoughts and desires. We can't make sense of living without others, and that includes other animals."[10] We must therefore acknowledge the challenges of these "complex relationships" rather than deny them, and thus, she argues, "we would do better to think about how to be more perceptive and more responsive to the deeply entangled relationships we are in."[11] This, she writes, "is the entanglement of entangled empathy. We are not just in relationships as selves with others, but our very selves are constituted by these relations."[12] I would add that these complex relationships not only involve our nonhuman animal kin, but that they may also incorporate places, ecosystems, plant life, and other terrestrial matter.

Gruen's concept of entangled empathy similarly suggests that our relationships with other animals constitute unbounded, entangled relationships, which take shape in ever-changing places and contexts. Such relationships may apply to how we conceptualize connectivity and coexistence when it comes to corridor planning and design. Gruen does caution, however, that we must keep our own projections in check as we consider the challenges and nuances of an empathetic approach in caring for nonhuman animals. For instance, Gruen notes that to genuinely empathize, we must "focus carefully on and take account of the specific context of the other, their idiosyncratic desires and personality, and the processes that shaped who they are."[13] This kind of position requires being open-minded to "learning and gathering information across differences," as well as a "commitment to critical reflection, and ideally consultation with people who have experience with and knowledge of the life-worlds of specific others."[14] Such openness to accounting for the specific contexts of nonhuman animals and their habitats can help work against bias or projection.

When we project, Gruen writes, the outcome tends to be less productive because projection can result in the needs of the other not being met in favor of what wittingly or unwittingly makes the caregiver more comfortable on some level. Gruen also notes the tendency for empathy to open the door to bias; thus, she emphasizes that we must not become complacent in our assumptions or potential biases around even the most compassionate and altruistic of interventions, and that "being able to answer questions about the specific other with whom one is empathizing can help minimize the dangers of projection as well as the various biases that tend to be associated with empathy."[15]

One example of the need to avoid potential bias would be, as Chapter 3 discussed, the earlier tendencies to design wildlife corridors around the needs of charismatic megafauna or popular flagship or keystone species. Rather, recent work in corridor ecology has shown that corridor design should account not only for flagship and keystone species but also for the needs of as many species in the habitat as possible, because, as described earlier, research has demonstrated that what serves as a corridor for one species may not necessarily work as well for another. On the one hand, the public's potential bias toward or identification with flagship species could have one unintended outcome or even perceived benefit of helping garner public support for wildlife corridor projects, and public support for such projects is necessary; on the other hand, however, attempting to account for the needs of *all* species has been shown to have better overall success in the long run and also better accounts for a compassionate conservation approach that avoids bias. That said, the aim of this book is not to suggest that every corridor project can or must apply a compassionate conservation approach in a neat and tidy package, for there is no perfect solution; rather, I suggest that proceeding with nuance and critical awareness can move the needle in a productive direction. When it comes to garnering public support for corridor projects, education campaigns are also beneficial, as I discuss later in the chapter about light pollution and migratory birds.

Conservation Projects and Different Ideas about Animal Rights

While it is likely that no single wildlife corridor project will enact a specific approach perfectly, Bekoff and Gruen begin to help us understand some of the nuances involved in understanding our relationships with and value systems related to our relationships with nonhuman animals. In doing so, Bekoff also describes broadly the positions of animal welfarists, animal rightists, utilitarians, and conservation biologists. The welfarist position would have that "while humans should not wantonly exploit animals, as long as we make animals' lives comfortable, physically and psychologically, we're respecting their welfare [...] In the end, welfarists agree that the pain and death animals suffer is sometimes justified because of the benefits that humans derive."[16] Animal rightists, on the other hand, emphasize that "animals' lives are valuable in and of themselves, not valuable just because of what they can do for humans or because they look or behave like us."[17] As Bekoff describes, animals "are not property or 'things,' but rather living organisms, subjects of a life" who are worthy of our compassion and respect.[18] Finally, Bekoff notes that the field of conservation biology has historically tended to favor the welfarist position; welfarists, he notes, are "willing to trade-off individuals' lives for the perceived good of higher levels of organization such as ecosystems, populations, or species."[19] Here he cites the example of reintroducing the Canadian lynx into Colorado, where there was not enough food for them in the habitat, causing many lynx to starve; in such a case, "some conservationists and environmentalists, in contrast to rightists, argued that the death of some individuals [...] was permissible for the perceived good of the species," whereas compassionate conservation would tend to look for an alternate solution.[20] Thus, how we perceive the lives of animals and our relationships with them and the environment can indeed give way to vastly different ideas and priorities about "who lives, who dies, and why."[21]

Critiques of Compassionate Conservation

Based on these ideas, as mentioned in the previous chapter, some critics of compassionate conservation have suggested that empathy and compassion are incompatible with the goals of conservation science. Researchers Griffin et al. argue that "empathy is subject to significant biases and that inflexible adherence to moral rules can result in a 'do nothing' approach."[22] As mentioned earlier, Gruen likewise cautions against projection and bias when it comes to how we care for nonhuman animals and argues that we should proceed with critical awareness about such pitfalls. With this, she notes that entangled empathy can lead more toward action than a "do nothing" approach because of its focus on "another's experiential wellbeing" and its tendency to lead to action based on an assessment of what seems to be the best, or most compassionate choice, in helping to pursue the wellbeing of another.[23] She also notes that projection can be avoided if we remain focused on the specific context of the other, and if we remain interested in learning about the contexts that shape their lives.

Griffin et al. also argue that "[o]f central concern to those who adhere to the ideas of compassionate conservation is the killing of introduced species as a means to restore and manage ecosystems."[24] They then argue that "faced with the moral dilemma of sacrificing a few to save the many, compassionate conservationists defer from harming anything and in so doing ultimately harm many more."[25] They suggest that "[c]ompassionate conservationists do not want to kill because it is morally wrong to harm, that is, adherence to the moral rule takes precedence over the utilitarian benefit of saving many."[26] The researchers argue that a "do nothing" approach is then preferred by compassionate conservationists over intervention, and that "compassionate conservationists have been vocal in expressing the view that humans should step back from managing the natural world, and 'let nature take over.'"[27] Citing two of the guiding principles of compassionate conservation—the goal of upholding a "do no harm" philosophy and the view that all individual animal lives matter, Griffin et al. note that, from the compassionate conservation vantage point, the "conservation practice of killing nonhuman animals violates legitimate values of life that place the emphasis on individual-level animal welfare."[28] They also suggest that advocates of compassionate conservation feel "that the utilitarian, evidence based decision-making frameworks that underpin conservation science have failed and should therefore be overhauled and replaced with empathy and the moral principles of 'first, do no harm' and 'individuals matter,' despite reviews finding conservation actions work."[29] Here, compassionate conservation and a "do no harm" philosophy may get conflated with the idea that such perspectives eschew a science-guided approach to conservation practice—a perspective that compassionate conservation would likely push back against.

Finding Middle Ground in the Nuances of Engagement and Critical Awareness

With these views in mind, I respectfully suggest that we need not necessarily swing between two ends of the pendulum, that is, culling or killing introduced or native species in order to balance out population numbers or restore ecosystems on the one hand, or doing nothing on the other hand. In fact, I suggest that wildlife corridors constitute an evidence-based, decision-making framework that implicitly supports solutions that represent a potential middle ground when it comes to conservation projects that aim to restore connectivity and preserve biodiversity. Moreover, as some of the illustrative cases in this book show, many of the stakeholders who speak on behalf of corridor projects implicitly and explicitly express a combination of values related to the need for wildlife corridors. Projects like the Y2Y corridor, for instance, have more explicitly sought to be inclusive of all beings, humans and nonhumans, who make use of that extensive ecosystem. Such projects have implicitly shown that it is possible to enact a compassionate conservation approach with corridor planning while circumventing more extreme measures of population control, for instance. As Bekoff has said of the Y2Y project: "This science-based project is all about connectivity. It involves maintaining and building natural corridors—replete with underpasses and overpasses around roads and creating protected areas near human communities—so that animals can move freely and safely without having to worry about people or cars."[30] Moreover, the Y2Y project "is also concerned with protecting native plants and allowing human communities to thrive. Thus, its goals are inclusive of all beings, nonhuman and human."[31]

Again, while there is seldom, if ever, a perfect solution, and while conservation management approaches like biodiversity offsetting may sometimes be necessary, as in the case of the Chiltern Railways reconstruction project discussed in this book, I suggest that an awareness of compassionate conservation and empathy provide a starting point that invites us to consider the nuance and middle ground when it comes to ecosystem management and connectivity restoration.

To this end, the concept of "rewilding" introduces a more relational perspective that allows us to proceed with critical awareness and respect in corridor planning. As Bekoff describes: "Rewilding personalizes what conservation projects try to accomplish in the world by building wildlife bridges and underpasses so that animals can move freely between fragmented areas. I see rewilding our hearts as a dynamic, intimate process that fosters corridors of coexistence and compassion for animals and their homes at the same time that it facilitates corridors in ourselves that connect our heart and brain, our caring and awareness."[32] In doing so, rewilding encourages a more relational perspective that acknowledges the need for ecosystem management in an Anthropocene era of human development, at the same time that it acknowledges the need for coexistence between people and wildlife. As he puts it: "Rewilding our hearts and rewilding the human dimension mean redefining the borders in our interactions with other animals and overcoming the cognitive dissonance that abounds globally."[33] Moreover, he writes: "I like to use the term 'borders' rather than 'boundaries' or 'barriers' because the latter words imply a less permeable interface between 'them' and 'us.' Redefining and softening these borders and distinctions is what rewilding is all about."[34]

In addition to these ideas about how we understand the lives of animals, Bekoff raises another important, overarching question about the larger motivations for and consequences of conservation biology—that built into its mission is the assumption that conservation biology, on some level, involves human intervention into the environment and often accounts for or attempts to course-correct prior human intervention in the first place. That is, conservation biology implicitly places humans at the forefront of efforts to protect species through means that, nonetheless, involve modes of human intervention and control, such as biodiversity offsetting, no matter how well-meaning those efforts might be. Here, Bekoff again calls into question human exceptionalism: "Can we really recreate or restore ecosystems? [...] Can we or should we try to 'do it all?'" Nonetheless, Bekoff is willing to acknowledge and even engage the paradox that necessarily underpins ideas about compassionate conservation. Occupying the mossy terrain that exists someplace between hope and despair, Bekoff argues that "since we decide who lives and who dies, compassionate conservation can easily be integrated into decisions about the fate of individual animals."[35] Bekoff is acutely aware of the role that humans play in managing the lives of nonhuman animals and wants to complicate these hegemonic binaries in ways that look for the potentially nuanced relationalities within them; doing so then requires that we engage more empathetically with the lives of vulnerable nonhuman animals. When it comes to countering human-centric views of the environment, as well as speciesism that values certain animal species above others, then, critical awareness and nuance are arguably necessary elements of conservation planning and understanding the value of ecosystems from a more ecologistic perspective. Here, TEK arguably aligns with and complements ideas about rewilding, coexistence, and empathy for species as it pertains to conceptualizing wildlife corridors.

Traditional Ecological Knowledge

There are several ways of understanding and defining TEK. As mentioned in the Introduction, TEK may be understood broadly as "a cumulative body of knowledge, practice, and belief, evolving by adaptive processes and handed down through generations by cultural transmission, about the relationship of living beings (including humans) with one another and with their environment."[36] Anishinaabe author and environmental activist Winona LaDuke further defines TEK as "the culturally and spiritually based way in which Indigenous people relate to their ecosystems. This knowledge is founded on spiritual-cultural instructions from time immemorial and on generations of careful observation within an ecosystem."[37] It is also important to acknowledge that the term "TEK" is, in and of itself, derived from within academia, and so it is worth noting that such lexicons do not necessarily apply in the same way across different contexts. As McGregor describes:

> The major difference between Aboriginal and non-Aboriginal Eurocentric views of TEK is that Aboriginal views of TEK are action oriented. One *does* TEK; it is not limited to a "body of knowledge." Non-Aboriginal views of TEK are more concerned with what the knowledge consists of and how it is transmitted. TEK is not just knowledge *about* the relationships with Creation, it *is* the relationship with Creation; it is the *way* that one relates.[38]

This understanding of TEK as an active *verb*, in a sense, or as a relational concept is arguably important for wildlife corridors and connectivity projects that shape land use and alter ecosystems and landscapes; it also plays a role in helping to understand the need for coexistence between people, landscapes, and wildlife. Stephanie Gillin, tribal member and wildlife biologist with the Confederated Salish and Kootenai Tribes, implicitly describes TEK as a relationship, and as a mode of compassionate conservation, when she discusses the need for coexistence between people and wildlife, as well as the need for the wildlife corridor on U.S. Highway 93 North that runs through the Flathead Indian Reservation in western Montana, discussed later in this book:

> [I]f you go back to our creation stories, we believe that animals were here before we were, and they helped prepare the Earth for us to be here. So in a way, it's kind of us being their voice, and us protecting them too, because they don't have a voice. So it's our generational duty, our responsibility as Tribal people to help protect wildlife because of our relationship with them.[39]

In this way, TEK is lived and practiced—not studied or observed. Kimmerer likewise understands TEK as action-based, and as focused on giving back to the land. As she suggests: "We need a different kind of science—and we have one. It has existed for millennia. It is called traditional ecological knowledge, which is rich in teachings about how people can give back to land."[40] McGregor similarly conveys: "The underlying aim of the science of ecology, therefore, the understanding of the web of relationships with the 'household' of Nature, is not modern science's sole property. *Understanding the relationship scientifically is not enough—living and nurturing these relationships is the key*. This is the ecology of the Native community."[41]

Finally, Ramos et al. acknowledge that there are a range of ways to conceptualize and define TEK depending on the context; thus, rather than trying to reach a singular consensus, they suggest that researchers should consider the role that TEK plays in collaborative relationships such as those involving federal agencies and tribal peoples. In this way, "TEK can be viewed as a collaborative concept that serves to unite diverse populations to continually learn from one another about philosophies of knowledge, how various approaches can be blended together to better steward natural resources and adapt to climate change."[42] Clearly, then, TEK can be understood from different vantage points and applied in a range of ways, either more implicitly or more explicitly, and through more qualitative or quantitative approaches, depending on the goals, audience, and context of a given conservation project.

When it comes to wildlife corridors, inclusivity toward and communication with all stakeholders and local peoples and communities has emerged as a critical component of land management and decision-making. Ongoing dialogue among all stakeholders is critical to the success and well-being of reconnected landscapes. As Haq et al. describe in their study of the woody tree species from the Dering-Dibru Saikhowa Elephant Corridor in northeast India: "Local ecological expertise can offer insightful opinions on sustainable forest management techniques that have evolved endogenously over many generations in the natural environment."[43] To incorporate TEK and practices into conservation projects is thus "crucial for the ecological transition since it may encourage sustainable land use practices, enhance biodiversity, and assist and empower local communities."[44] Moreover, as Haq et al. note, integrating TEK can help in "adaptive management given that it frequently supplements previously gathered ecological data by providing additional information at a finer spatial scale than scientific data," which provides an additional, valuable window into ecosystem dynamics and their connections with "societal values, activities, and resource use patterns."[45] In recent years, TEK has become more widely incorporated into conservation planning, outside of the often narrower confines of academic research.

On November 15, 2021, the White House Office of Science and Technology Policy (OSTP) and the White House Council on Environmental Quality (CEQ) issued a joint memo that recognized TEK—what they referred to as Indigenous Traditional Ecological Knowledge (ITEK) "as one of the many important bodies of knowledge that contributes to the scientific, technical, social, and economic advancements of the United States and to our collective understanding of the natural world."[46] The memo acknowledged that ITEK:

> has evolved over millennia, continues to evolve, and includes insights based on evidence acquired through direct contact with the environment and long-term experiences, as well as extensive observations, lessons, and skills passed from generation to generation. ITEK is owned by Indigenous people—including, but not limited to, Tribal Nations, Native Americans, Alaska Natives, and Native Hawaiians.[47]

Based on requests for guidance on how to create partnerships between federal agencies and tribal Nations, the Office of the President recommended that "the Federal Government should engage with ITEK only through relationships with Tribal Nations and Native communities and in a manner that respects the rights of knowledge holders to control access to their knowledge, to grant or withhold permission, and to dictate the terms of its application."[48] The memo stated that, where appropriate, TEK should "inform Federal decision making along with scientific inquiry," and noted some of the ways in which Tribal Nations and Native communities are already collaborating with federal agencies to incorporate TEK into "knowledge- and evidence-based Federal Government decision making."[49] To this end, in 2021, the OSTP and CEQ formed an "Interagency Working Group on Indigenous Traditional Ecological Knowledge" whose charge will be, in part, to develop and provide "best practices on how to collaborate with Tribal Nations and Native communities" and apply TEK in ways that achieve "mutually beneficial outcomes," including how to navigate federal laws and processes, "and how to appropriately respect the knowledge holders' rights to decline participation in efforts to collaborate."[50] Finally, the memo included an addendum that highlighted some past and current examples of collaborations between Native communities and the federal government.

One such example involves the recent creation of the Papahānaumokuākea Marine National Monument (PMNM) in the Northwestern Hawaiian Islands, which is discussed in more detail later in this book. The PMNM has a unique co-management structure and is overseen collaboratively by the "National Oceanic and Atmospheric Administration, U.S. Fish and Wildlife Service, the State of Hawai'i Office of Hawaiian Affairs, and the Hawai'i

Department of Land and Natural Resources, and Native Hawaiians have consistently led the development and governance of the monument."[51] Its management is based on traditional Hawaiian knowledge practices, which understand cultural heritage and the natural world as synonymous with one another. In doing so, the PMNM also implicitly takes a compassionate conservation approach to ecosystem management.

Conclusion

Shortly following the publication of the presidential memo, a statement published by United South and Eastern Tribes, Inc. (USET) acknowledged that the presidential memo "represents a significant advancement by this [the Biden] administration in deepening our nation-to-nation relationships. [...] and serves as an example of substantive and meaningful Indian Country inclusion and visibility."[52] The USET statement likewise added that "[w]hile this announcement alone does not correct our larger invisibility challenges, it represents an acknowledgment that we co-exist as sovereigns and therefore, should have a vested interest in the lands we share, including working in partnership to face our mutual challenges, such as the protection and stewardship of our lands and environment."[53]

As the role of TEK in conservation management and planning continues to gain long overdue support across organizations and institutions, its vital role in connectivity projects becomes increasingly clear; moreover, we may see the ways that TEK is aligned with a compassionate conservation approach that is inclusive of all beings, nonhuman and human, and that values coexistence among people, wildlife, and the landscapes we co-inhabit. While not every corridor or connectivity project may leverage TEK in the same way or to the same extent, it is clear that TEK is a valuable source of cultural and ecological knowledge and ought to be incorporated into connectivity projects when it is relevant and feasible to do so.

In the illustrative cases that follow, this book explores some specific examples of how wildlife corridors may function in sync with these ideas and philosophies—again, in different ways and to different extents, and how such work can benefit landscapes, wildlife, and people. The cases also help demonstrate how we can incorporate connectivity at a range of scales, through examples of both larger- and smaller-scale projects, all of which work to promote a culture of connectivity and coexistence in different ways.

Notes

1 UTS, "What Is."
2 Tobias, "Compassionate Conservation."

3 Tobias, "Compassionate Conservation."
4 Bekoff, Animal Manifesto, 40.
5 Bekoff, *Rewilding*, 13.
6 Bekoff, *Rewilding*, 9.
7 Bekoff, *Rewilding*, 9.
8 Gruen, *Entangled Empathy*, 50.
9 Gruen, *Entangled Empathy*, 51.
10 Gruen, *Entangled Empathy*, 64.
11 Gruen, *Entangled Empathy*, 63–64.
12 Gruen, *Entangled Empathy*, 64.
13 Gruen, *Entangled Empathy*, 60.
14 Gruen, *Entangled Empathy*, 60.
15 Gruen, *Entangled Empathy*, 60.
16 Tobias, "Compassionate Conservation."
17 Tobias, "Compassionate Conservation."
18 Tobias, "Compassionate Conservation."
19 Tobias, "Compassionate Conservation."
20 Tobias, "Compassionate Conservation."
21 Tobias, "Compassionate Conservation."
22 Griffin et al., "Compassionate Conservation Clashes with Conservation Biology," 1.
23 Gruen, *Entangled Empathy*, 51.
24 Griffin et al., "Compassionate Conservation Clashes with Conservation Biology," 4.
25 Griffin et al., "Compassionate Conservation Clashes with Conservation Biology," 5.
26 Griffin et al., "Compassionate Conservation Clashes with Conservation Biology," 5.
27 Griffin et al., "Compassionate Conservation Clashes with Conservation Biology," 5.
28 Griffin et al., "Compassionate Conservation Clashes with Conservation Biology," 4–5.
29 Griffin et al., "Compassionate Conservation Clashes with Conservation Biology," 6.
30 Bekoff, *Rewilding*, 9.
31 Bekoff, *Rewilding*, 9.
32 Bekoff, *Rewilding*, 12.
33 Bekoff, *Rewilding*, 12.
34 Bekoff, *Rewilding*, 13.
35 Tobias, "Compassionate Conservation."
36 Berkes, *Sacred Ecology*, 7.
37 McGregor, "Coming Full Circle," 393–394.
38 McGregor, "Coming Full Circle," 394.
39 Christy and DiGirolamo, "Wildlife Crossings Built with Tribal Knowledge Drastically Reduce Collisions," 2022.
40 Planet Forward Staff, "2017 Summit."
41 Cajete 95; quoted in McGregor, "Coming Full Circle," 394.
42 Ramos et al., "Introduction to Traditional Ecological Knowledge in Wildlife Conservation," 11.
43 Haq et al., "Integrating Traditional Ecological Knowledge into Habitat Restoration," 16–17.
44 Haq et al., "Integrating Traditional Ecological Knowledge into Habitat Restoration," 17.
45 Haq et al., "Integrating Traditional Ecological Knowledge into Habitat Restoration," 2.
46 Executive Office of the President, "Memorandum."
47 Executive Office of the President, "Memorandum."

48 Executive Office of the President, "Memorandum."
49 Executive Office of the President, "Memorandum."
50 Executive Office of the President, "Memorandum."
51 Executive Office of the President, "Memorandum."
52 Francis, "Honoring Traditional Ecological Knowledge Is Critical."
53 Francis, "Honoring Traditional Ecological Knowledge Is Critical."

Chapter 5

THE WILDLIFE CROSSING ON THE FLATHEAD INDIAN RESERVATION IN MONTANA, USA

Respecting the Spirit of Place

The section of U.S. Highway 93 North that runs through the Flathead Indian Reservation in western Montana spans approximately 56 miles, from about Evaro, MT, in the south, to Polson, MT, in the north. This north-south stretch of U.S. 93 Highway cuts right through one of Montana's primary east-west wildlife migration corridors and, in doing so, "runs through large expanses of wildlife habitat, including the Mission Mountains, the Bob Marshall Wilderness Complex, and the Selway-Bitterroot Wilderness Area," thus fragmenting these habitats and making improved connectivity a priority in this area.[1] Recent infrastructure and corridor-related work along this stretch of highway has helped to restore connectivity in this area and has benefited ecosystems, people, and wildlife in the process.

Just outside of the Evaro area, near Arlee, along Highway 93, travelers will come across what is known as the "Animals' Trail," or a 197-foot-wide vegetated bridge, which allows wildlife to safely cross over the highway. The overpass itself is most visible along this stretch of highway; however, it is but one of many wildlife crossings along Highway 93. In short, the wildlife-friendly structures in this area represent the joint efforts of the Montana Department of Transportation (MDT), Confederated Salish and Kootenai Tribes (CSKT), and Federal Highways Administration (FHWA), which worked together to build this infrastructure not only to reduce human-wildlife traffic accidents but also to protect wildlife migration routes. The project was first proposed in 1989 by the MDT, which wanted to expand this section of U.S. 93 into a four-lane highway. The expansion, however, would have extended into the Flathead Indian Reservation, home to the CSKT. Eventually, the FHWA, MDT, and CSKT convened and "established a tri-governmental team to reach an agreement. From that process came a radical idea: instead of focusing on how the road will impact the land, focus on how the land should shape the road. The team called this approach a 'Spirit of Place.'"[2] These groups worked together productively to improve connectivity in ways that not only increased human safety but were also sensitive to the needs of wildlife, in particular by incorporating traditional ecological knowledge (TEK) and respect for the land, through its focus on the "spirit of place," which "takes into account the surrounding mountains, plains, hills, forests, valleys, and sky. To focus on the spirit of place means incorporating knowledge and respect for the paths of waters, glaciers, winds, plants, animals, and native people—the whole continuum of what is seen, touched, felt, and traveled through."[3]

Notably, the design of the road is "premised on the idea that the road is a visitor and should respond to and respect the Spirit of Place."[4] The overarching philosophy for the reconstruction of the highway was to "protect cultural, aesthetic, recreational, and natural resources located along the highway corridor and to communicate the respect and value that is commonly held for these resources pursuant to traditional ways of the Tribes."[5] To date, the U.S. Highway 93 North connectivity project on the Flathead Indian Reservation represents one of the most extensive wildlife-centric highway re-connectivity efforts in North America.

Completing this connectivity project entailed reconstructing this 56-mile-long section of road and installing wildlife crossing structures at "39 locations and approximately 8.71 miles (14.01 km) of road with wildlife exclusion fences on both sides."[6] The goal of the mitigation measures was not only to improve driver safety by reducing wildlife-vehicle collisions but also to allow wildlife to continue to cross the road.[7] In particular, structural connectivity efforts along U.S. 93 North included the installation of wildlife fences and wildlife crossing structures along specific sections of road, and the subsequent monitoring of those structures to evaluate their effectiveness.[8] These wildlife fences are meant to prevent wildlife from entering the highway and instead guide them toward places that are safe to cross, thus avoiding collisions with vehicles. In addition to fences, underpasses, wildlife guards, and wildlife jump-outs were also installed (Figures 5.1, 5.2, 5.3, and 5.4).

The goals of increasing safety and providing new means of connectivity have been met, largely through the collective research efforts of several groups. Since about 2008, researchers at Montana State University have monitored usage of this stretch of U.S. Highway 93, and in recent years, local transportation authorities have begun to shift their mindsets about the ways that infrastructure like highway systems ought to be conceptualized and designed.[9] The resulting improvements in connectivity illustrate how corridor projects can not only incorporate the needs of various stakeholders

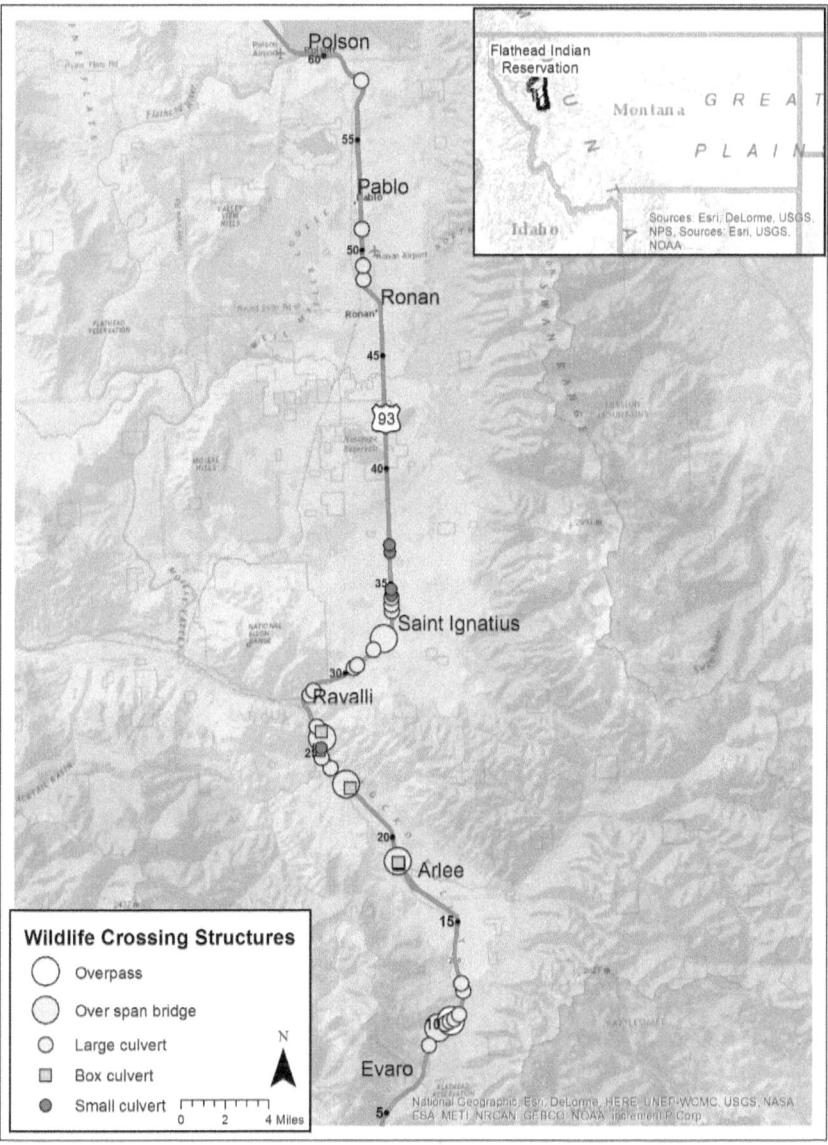

Figure 5.1 Map: Locations of wildlife crossing structures along U.S. Hwy 93, Montana, USA. Reprinted with permission from WTI-MSU.

but also work from a vantage point of incorporating TEK and compassionate conservation that accounts for the needs of human communities, land, and wildlife. As a research report from the Western Transportation Institute notes, "Values related to culture, landscape, and natural resources are not uniquely Native American. These values are present in almost any

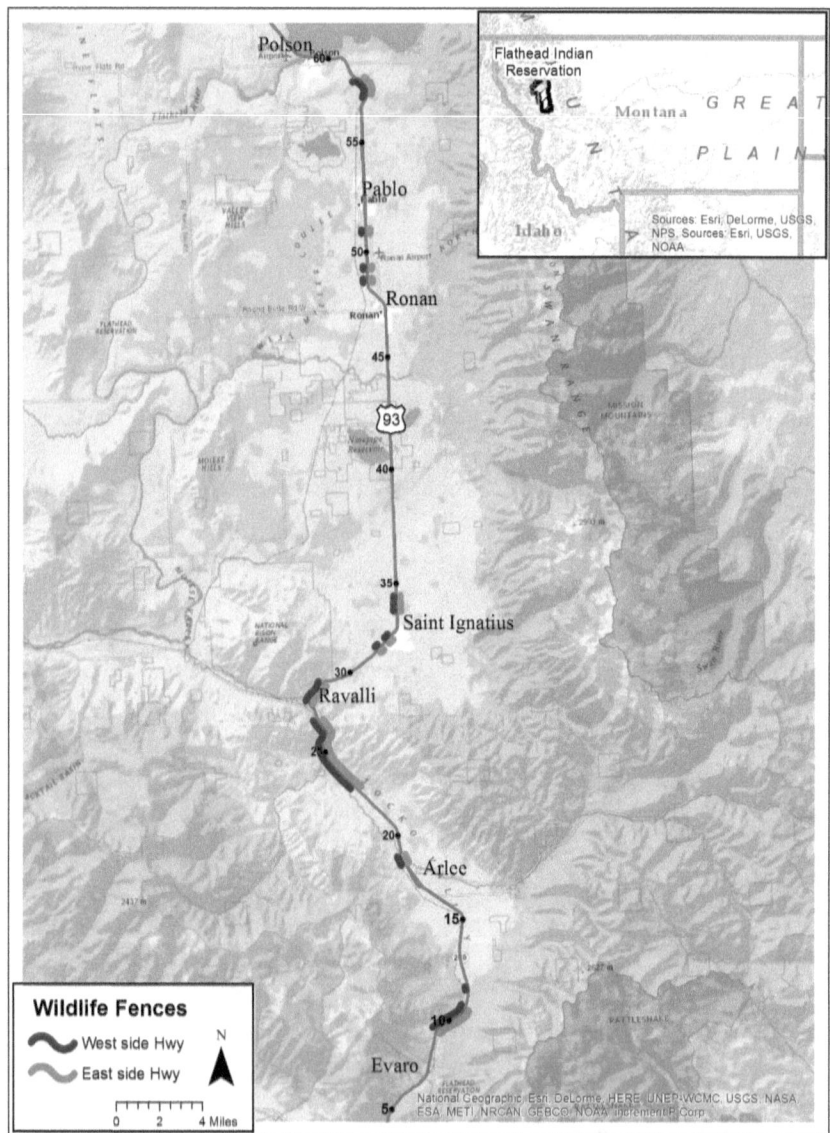

Figure 5.2 Map: Locations of wildlife exclusion fences along U.S. Hwy 93, Montana, USA. Reprinted with permission from WTI-MSU.

society. However, in the specific context of the reconstruction of a highway on a Native American reservation, these values were actually made an integral component of a context sensitive approach to redesigning a highway."[10] Such an approach is also consistent with the tenets of compassionate conservation, for as Marc Bekoff writes, "We must understand what

Figure 5.3 Wildlife overpass, or "Animals' Bridge," U.S. Hwy 93, Flathead Indian Reservation, Montana, USA. Copyright Marcel Huijser.

Figure 5.4 Wildlife jump-out or escape ramp along U.S. Hwy 93, Montana, USA. Copyright Marcel Huijser.

other individuals, human and nonhuman alike, want, need, and feel so that the collective is cared for, not just one species at the expense of countless others."[11]

Wildlife Fences Afford Both Structural and Functional Connectivity

Research to monitor the effects of wildlife fences after the fences were installed showed a more than 80 percent reduction in vehicle collisions with large mammals "if the fences and associated measures were installed over road lengths of at least 3.1 mi (5 km)."[12] Notably, if the wildlife fences were placed at shorter road lengths of under three miles long, their effectiveness dropped to about 50 percent and became less predictable in their positive impacts. As a report from the Western Transportation Institute describes: "The reduced effectiveness of short fenced road sections was related to fence end effects that resulted in a concentration of collisions at and near fence ends."[13] While fence end treatments were shown to improve the effectiveness of short-fenced road sections, research should also explore whether wildlife fences "can be extended to cover at least 3.1 mi (5 km) of continuous road length to reduce the fence end effects."[14] The effectiveness of wildlife fences based on the correlation between fence length and road length is "new knowledge" that was based in part on the results of the U.S. 93 North research project. The researchers note that this new knowledge is yet another gauge of success; that is, to gain knowledge then allows for new "policies and practices of highway and wildlife management agencies."[15]

Research was also conducted to assess the effectiveness of the highway reconstruction in reducing vehicle collisions with black bears and grizzly bears. It did not appear that the mitigation measures had as much of an impact on bear populations; this was most likely due to "the relatively short road lengths equipped with mitigation measures, the design of the wildlife fence, and the gaps in the wildlife fence at access roads and steep slopes."[16] Additionally, data revealed a "learning curve" for species like deer and black bears, who appeared to use the structures more over time: "While deer and black bear use can be considered high one year after construction, both species showed an increase in successful crossings for at least five years after construction. This suggests that wildlife use, specifically by deer and black bear, is likely to continue to grow."[17] Rob Ament, the Road Ecology program manager at Montana State University's Western Transportation Institute (WTI), noted that "[s]ome animals are initially wary of any artificial crossing structure. Research cameras have recorded animals approaching tunnels and bridges, then running away at the sound of a vehicle."[18] Over time, however, they become more accustomed to the structures, begin to use them more, and teach their young ones to use them as well.

Following the highway reconstruction project, deer highway crossings were either similar or increased, while black bear crossings remained similar. Deer population size did not increase after reconstruction, which led researchers to believe that the new construction did not negatively impact habitat connectivity for deer; rather, when taking into account the learning curve for deer, habitat connectivity actually increased in the reconstructed sections of the road.[19]

Many large mammals used the underpasses that were created, and their usage varied greatly—this was apparently independent of the fence length associated with the underpasses. In other words, there does not appear to be a correlation between fence length and underpass usage, which leads researchers to believe that larger mammals are influenced by other factors such as "the location of the structure in relation to the surrounding habitat, wildlife population density, and wildlife movements."[20] Nonetheless, the researchers note that even though "wildlife use of underpasses is highly variable […] an individual underpass may still have higher wildlife use if that underpass is connected to wildlife fences and if the fence length is long rather than short."[21]

Monitoring and Usage

After the U.S. Highway 93 connectivity project was complete, researchers installed wildlife cameras at 29 structures to record wildlife use. Notably, the cameras recorded "95,274 successful crossings in total or 22,648 successful crossings per year, which can be described as substantial."[22] White-tailed deer used the structures the most, with Mule deer and domestic dogs and cats following suit; additionally, black bears represented 1,531 successful crossings.[23] Overall, 20 different species of medium and large mammals used these structures. The researchers note that White-tailed deer tended to use "bridges, overpasses and large culverts more than expected. […] Mule deer also used bridges and large culverts more than expected. […] Black bear used a wider variety of structures, bridges, large culverts and small culverts, more than expected. Grizzly bears exclusively used large culverts. […] Elk and moose mostly or exclusively used the wildlife overpass."[24] The highway reconstruction project also implemented the use of wildlife guards and wildlife jump-outs.

Wildlife guards are similar to cattle guards and were installed at selected access roads along the highway, primarily to discourage ungulates "from accessing the fenced road corridor at access roads."[25] These guards were very successful in preventing deer from accessing the fenced road corridor but were not as successful at deterring other mammals.[26] Wildlife jump-outs are "earthen ramps within the fenced right-of-way"; if wildlife become caught between fences, then the jump-outs allow them to walk up an incline at the fence line and jump down to the other side of the fence, away from the highway.[27] The researchers note that jump-outs should ideally be low enough that wildlife can easily jump down to safety but high enough that they will not "jump into the fenced road corridor where they may be hit by vehicles."[28] Of course, the optimal height would then depend on the species for which it would be designed and their ability to jump in either direction, and the researchers suggest that creating such guidelines for specific species requires further research and monitoring.

Stakeholders, Governance, and the Value of Partnerships

The relative success of these reconstruction and connectivity efforts along U.S. Highway 93 North was due largely to the collective efforts of federal, state, and tribal governments. These groups not only agreed upon what mitigation measures would be implemented in the first place, but they also agreed upon what would then count as success. Encouragingly, nearly all of these measures of success were met. Most successful was the increased habitat connectivity for deer and black bears, and the strong function of the wildlife crossing structures in allowing for such increased connectivity.

The success of Montana's U.S. Highway 93 reconstruction has demonstrated the value of partnerships for the progress of wildlife corridors and crossings. These partnerships included "coalitions of nonprofits, land trusts, hunter and angler groups, local businesses, and counties," all of whom worked together to help reconstruct this portion of U.S. Highway 93. Moreover, an organization called Vital Ground, "a foundation dedicated to grizzly bear conservation in the Northern Rockies, has purchased Montana land parcels near some road crossings, connecting grizzly bear habitat that was once fragmented."[29] In addition to official organizations, local communities and community members have also played a role by communicating with researchers and state officials about the locations where animals are most frequently seen or hit by vehicles.

"The Road Is a Visitor"[30]

Improvements to this section of U.S. Highway 93 North perform many of the tenets of good wildlife corridor design. In this case, it is possible to see how strong efforts at structural connectivity can, in turn, support high functional connectivity. Again, functional connectivity also implicitly aligns more with a compassionate conservation and entangled empathy perspective because it seeks to understand and support connectivity from the perspective of individual species of wildlife. Moreover, effective biodiversity conservation requires that corridors simultaneously meet the needs of multiple species—not just large carnivores; the reconstruction of U.S. Highway 93 indeed meets this measure, thus aligning with approaches to entangled empathy and compassionate conservation, which would suggest that *every* species matters when it comes to conservation efforts. As Bekoff has noted: "The concept of rewilding is grounded in the premise that caring is okay. In fact, it is more than okay; it is essential. It is all right to imagine the perspective of nonhuman animals in order to take their well-being into account. [...] This means many things [...] but primarily it means opening our hearts and minds to others."[31] Finally, the highway reconstruction project has helped to increase habitat connectivity, especially for deer species, in the mitigated road sections. Such measures ultimately help preserve biodiversity and protect the well-being of species by encouraging the successful dispersal of individuals between populations to increase gene flow and long-term population viability.

With this specific example of the wildlife crossing on the Flathead Indian Reservation in Montana, however, what also stands out is the strong partnership of multiple governments and organizations and the need to respect the land—a need that was illuminated by Indigenous communities and viewed with consensus by all parties involved.

As Whisper Camel Means, tribal member and wildlife biologist with the CSKT, says:

> As tribal people living on the remnant of our homelands, our connection to our past, our connection to our culture, really rests in the fact that wild animals are still here—that we can use them for subsistence—that we can use them for cultural connection—that we need to have them here to have a healthy ecosystem and environment.[32]

Here, she also acknowledges the desire and ability of the Tribes to help make this project a reality: "One of the unique characteristics of this place is that the Tribes have the political will and position to be able to make this highway into what it is right now."[33]

The Highway 93 reconstruction project on the Flathead Indian Reservation illustrates what is possible when a shift in mindset, coupled with the cooperation and strong partnerships of governments, organizations, and communities comes together. From these more clearly articulated understandings and expressions of green infrastructure comes a mindset of respect for the land and the coexistence of humans and wildlife: "This transition in ideology was only able to occur when humans started to view the highway infrastructure as a part of the economy of nature rather than opposed as its antithesis."[34] In this way, "the highway environment has successfully started to be viewed as a diverse and dynamic network of complex fragmented patches and corridors to aid in the preservation of biodiversity, not its further destruction."[35] Additionally, to help protect wildlife, it "in turn helps the Native American tribes who have resided upon the same land for centuries, as the land and animals are considered and respectfully appreciated as natural and cultural resources for the Tribes."[36] As Germaine White, tribal member and education specialist with the CSKT puts it: "This is our homeland. And this has been our homeland from the very beginning of time. And it will be our homeland until the very end of time."[37] And as Tony Incashola, former director of the Selis Qlispe Culture Committee, describes, the improvements to U.S. Highway 93 not only "helped the Tribes in preserving a way of life for their people, but also preserving the animals that are part of our values."[38] This inclusive approach helps foreground an important component of connectivity projects; that is, similar to the Y2Y Conservation Initiative, the connectivity project at the Flathead Indian Reservation has demonstrated the importance of including the voices of Indigenous peoples and TEK in the management of conservation areas, which in turn supports coexistence and compassionate conservation.

Notes

1 Shinn, "Montana's Wildlife."
2 People's Way Partnership, "US 93 North."
3 People's Way Partnership, "US 93 North."
4 People's Way Partnership, "US 93 North."
5 Huijser et al., "US 93 North Post-Construction Wildlife-Vehicle Collision," 1.
6 Huijser et al., "US 93 North Post-Construction Wildlife-Vehicle Collision," 1.
7 Huijser et al., "US 93 North Post-Construction Wildlife-Vehicle Collision," 1.
8 Huijser et al., "US 93 North Post-Construction Wildlife-Vehicle Collision," 1.
9 Chaney, "Research on Highway 93."
10 Huijser et al., "US 93 North Post-Construction Wildlife-Vehicle Collision," 97.
11 Bekoff, *Rewilding*, 47.
12 Huijser et al., "US 93 North Post-Construction Wildlife-Vehicle Collision," 2.
13 Huijser et al., "US 93 North Post-Construction Wildlife-Vehicle Collision," 2.
14 Huijser et al., "US 93 North Post-Construction Wildlife-Vehicle Collision," 2.
15 Huijser et al., "US 93 North Post-Construction Wildlife-Vehicle Collision," 5.
16 Huijser et al., "US 93 North Post-Construction Wildlife-Vehicle Collision," 1.
17 Huijser et al., "US 93 North Post-Construction Wildlife-Vehicle Collision," 65.
18 Shinn, "Montana's Wildlife."
19 Huijser et al., "US 93 North Post-Construction Wildlife-Vehicle Collision," 3. For further discussion of wildlife-vehicle accidents related to deer specifically, see Huijser and McGowen, "Reducing Wildlife-Vehicle Collisions."

20 Huijser et al., "US 93 North Post-Construction Wildlife-Vehicle Collision," 3.
21 Huijser et al., "US 93 North Post-Construction Wildlife-Vehicle Collision," 3.
22 Huijser et al., "US 93 North Post-Construction Wildlife-Vehicle Collision," 65.
23 Huijser et al., "US 93 North Post-Construction Wildlife-Vehicle Collision," 65.
24 Huijser et al., "US 93 North Post-Construction Wildlife-Vehicle Collision," 65.
25 Huijser et al., "US 93 North Post-Construction Wildlife-Vehicle Collision," 3.
26 Huijser et al. "US 93 North Post-Construction Wildlife-Vehicle Collision," 3.
27 Huijser et al., "US 93 North Post-Construction Wildlife-Vehicle Collision," 4.
28 Huijser et al., "US 93 North Post-Construction Wildlife-Vehicle Collision," 4.
29 Shinn, "Montana's Wildlife."
30 People's Way Partnership, "US 93 North."
31 Bekoff, *Rewilding*, 5.
32 Christy and DiGirolamo, "Wildlife Crossings Built with Tribal Knowledge."
33 Christy and DiGirolamo, "Wildlife Crossings Built with Tribal Knowledge."
34 Li-Chee-Ming, "Preserving Montana's Biodiversity."
35 Li-Chee-Ming, "Preserving Montana's Biodiversity."
36 Li-Chee-Ming, "Preserving Montana's Biodiversity."
37 Christy and DiGirolamo, "Wildlife Crossings Built with Tribal Knowledge."
38 Christy and DiGirolamo, "Wildlife Crossings Built with Tribal Knowledge."

Chapter 6

THE MONKTON WILDLIFE CROSSING AND THE BLUE-SPOTTED SALAMANDER

Vermont's First Amphibian Crossing Tunnels

Monkton Road is an approximately 10-mile stretch of road in Northwest Vermont, about 20 miles south of Burlington. For decades, it was a lightly traveled, local road, and traffic was not much of a concern for people or wildlife. In fact, Monkton Road was actually a rural dirt road for over a century until increased development prompted the need to pave the road. As Vermont's population grew, so did the number of vehicles on this road, also in part because GPS devices offered the route as a shortcut to Burlington. Soon, the rate of annual average daily traffic on the road reached between 2,000 and 3,000 vehicles, which became a threat to local amphibian populations who must cross this road every spring in order to breed.[1]

Or, as Chris Slesar, environmental resources coordinator at the Vermont Agency of Transportation (VTrans), put it more empathetically, 2,000–3,000 vehicles a day "is not a terribly busy road by US standards. However, it is unquestionably and significantly busy if your lifecycle requires you to crawl slowly across the road on your belly in the dark, sans reflective apparel, at least twice a year."[2] That is, this relatively short section of Monkton Road, just under a mile long, bisects two important pieces of amphibian habitat. Monkton Road runs directly between upland and wetland habitats, which means amphibians must cross this road to breed during their spring migration.[3]

For many years, a greater number of amphibians were being reproduced than were being killed on this stretch of road. More recently, however, increased traffic was claiming the lives of about 50 percent of amphibians who tried to cross this section of the road during their spring migration. Groups of concerned citizens and local researchers would gather at night to help move these amphibians across the road, attempting "nocturnal bucket-brigade rescues for the amphibians," which was limited in its success and also put people at risk.[4] In 2005, it became clear that this grassroots method, in which citizens manually carried these amphibians across the road, was not sustainable in the long term.[5]

When it became clear that grassroots methods to assist these creatures in crossing the road were helpful for these amphibians but not a fully safe or sustainable long-term solution, local wildlife biologists began a research project to help make the case that an infrastructure fix was needed at this site. The data collected in association with this research was ultimately integral to securing two federal grants that would help fund the wildlife crossings that were eventually constructed at this site.[6] In this way, evidence-based approaches align with a compassionate mindset in an effort to foster coexistence.

While human populations use the Monkton-Vergennes Road for commuting and recreation purposes, wildlife must also use this road to move between their upland wintering habitat and spring breeding wetlands. The blue-spotted salamander is among the diverse group of amphibians that live in this area and must navigate this busy stretch of road. This species of salamander is considered "a species of special concern" in Vermont[7] (Figure 6.1).

The blue-spotted salamander *(Ambystoma laterale)* is the smallest of the mole salamander species that resides in Vermont. It can grow up to five inches long and is "black with small light blue spots on their backs and sides."[8] With the

Figure 6.1 Blue-spotted salamander. Courtesy of the Vermont Agency of Transportation.

exception of the Redback salamander, all salamanders in Vermont live in the water for some of their lives. Salamanders generally seek out "damp, cool places during the day and only venture out in the cooler night air."[9] They require "the moisture found in wetlands and on the cool forest floor to keep their skin wet to prevent them from drying out" and have a complex courtship ritual that is ecosystem dependent.[10]

Throughout late March and early April, the male and female spotted salamanders migrate to vernal pools to breed. The male drops the sperm packet in shallow water; the female then takes up the spermatophores and fertilizes the eggs internally. In a few days, the female will lay approximately 150 eggs at one time. These eggs are "attached to vegetation in shallow water," and a single female can lay "two or three egg masses," which usually hatch within a couple of months.[11] Again, a portion of Monkton Road relevant to the salamanders' mating habits bisects the upland and wetland ecosystems that the salamanders require for mating and laying their eggs, and so these amphibians must cross the road to breed during their spring migration. The Monkton Wildlife Crossing Project took shape when concerned citizens saw that this population of amphibians in particular was threatened by the increasing traffic and development.[12]

Efforts to retrofit the section of Monkton Road entailed a coordinated effort that combined data collection with fundraising, public meetings, and engineering plans.[13] The wildlife crossing tunnels were completed in the fall of 2016, and the "$290,000 project took a decade of planning and fundraising by the Monkton Conservation Commission and the Lewis Creek Association."[14] The research began in earnest back in 1997, when Jim Andrews, coordinator of the Vermont Reptile and Amphibian Atlas Project, and Monkton resident and wildlife biologist Steve Parren realized that "the combination of upland hardwood habitat adjacent to a spectacular wetland complex, known locally as the Huizenga Swamp [...] in the Champlain Valley of Vermont, was the perfect mix to provide everything a cornucopia of amphibians could possibly need to thrive for centuries."[15] These amphibians winter underground in upland forest, but in early spring, when Vermont's "mud season" begins and the ice on the wetlands starts to melt, these amphibians emerge from underground and make their way to the wetlands to breed.

During the spring of 1997, Parren started making visits to the site and collected data about the amphibian movement he observed; he monitored the "date, time, weather conditions, number of animals, species, direction travelling, number of cars passing, and successful crossing vs. roadkill."[16] This data soon revealed a long list of amphibians who were regularly crossing this stretch of road, including: spotted salamander *(Ambystoma maculatum)*, blue-spotted salamander *(Ambystoma laterale)*, blue-spotted/jefferson salamander hybrid group *(Ambystoma jeffersonianum x laterale complex)*, four-toed salamander *(Hemidactylium scutatum)*, eastern newt *(Notophthalmus viridescens)*, spring peeper, *(Pseudacris crucifer)*, Wood Frog *(Lithobates sylvaticus)*, American toad *(Anaxyrus americanus)*, leopard frog *(Lithobates pipiens)*, green frog *(Lithobates clamitans)*, and gray tree frog *(Hyla versicolor)*.[17]

The Wildlife Crossing Structures: Leveraging Existing Landscape to Benefit Multiple Species

Ecologists, engineers, and grassroots organizations agreed that the only long-term solution to ensure the sustainability of this population of amphibians and increase their survival rate was to reconstruct the existing road with wildlife crossing tunnels.[18] In essence, the crossings consist of "two large concrete culverts with a natural bottom. Concrete walls will guide and funnel the amphibians into the culverts during their spring and fall migrations."[19] To help guide amphibians toward these tunnels, 200-foot-long "wing-walls" were constructed that served as funnels that would then "direct amphibians into the five-foot-wide tunnels under the road. When the animals return to the uplands after breeding and egg-laying, wing-walls on the downhill side funnel them into the tunnels again."[20] The greatest challenge in achieving structural connectivity was designing cost-effective retaining walls that would also be low maintenance.[21] Stackable concrete blocks were deemed the best solution and were used to build the walls (Figure 6.2)

Much like the effectiveness of the wildlife fences along U.S. Highway 93 North that run through the Flathead Indian Reservation in Montana, the success of the retaining walls, as gauged by the number of animals that they help guide through the crossings, is directly connected to the length of the walls. The retaining walls were constructed to keep the amphibians off the road and direct them to the culverts, and these walls turned out to be an "indispensable" and critical aspect in the wildlife crossing's design; that is, without retaining walls, also called "drift fencing," it would have been "utterly dumb luck for the animals to find the culverts."[22]

The culverts were designed as "5' x 5' pre-cast concrete bottomless box culverts."[23] They adhered to general wildlife corridor best practices for design in that they leveraged the existing landscape because the "boxes are buried into the natural substrate that follows the topographical contours of the landform."[24] Next, in an effort to facilitate airflow and moisture into the culverts, "two slotted grates" that resemble manhole covers were "installed in the center of each lane at each crossing structure."[25] The culverts are also large enough that ambient light can pass through them; light and moisture can enter through grates at the top, which is "critical in aiding wildlife through

Figure 6.2 Monkton wildlife crossing tunnel. Courtesy of the Vermont Agency of Transportation.

the structures."[26] Finally, square pieces of slate were "placed along the retaining walls and within the culverts as cover objects" or small platforms, such that migrating amphibians could seek shelter there during their migration season.[27]

Similar to the underpasses that run beneath the section of Florida's State 80 highway to protect the Florida panther, the Vermont wildlife crossing was also designed with a "flagship species" in mind; that is, the crossing tunnels were primarily constructed to help the amphibians that enter and exit the Huizenga Swamp, with the spotted salamander being among the amphibians who were considered most vulnerable. And as an added benefit to biodiversity preservation, the wildlife tunnels were also designed to be large enough not only to accommodate amphibians but also to allow larger species to safely cross the road.

Conclusion: Wildlife Crossing Tunnels as Acts of Compassionate Conservation

During the first spring migration season that the wildlife crossing was in use, cameras recorded over 2,000 amphibians using the structures. In recent years, as the site continues to be monitored, the numbers have been shown to be similar, with an average of 2,000 amphibians crossing each year. In addition, the wildlife crossing continues to help other animals as well; cameras have captured photos of "bobcat, mink, Virginia opossum, porcupine, ermine, Milk Snakes, Garter Snakes, groundhog, eastern cotton tail, raccoon, and mice sharing the space with amphibians."[28] The Monkton Wildlife Crossing Tunnels demonstrate not only how an interdisciplinary group along with concerned citizens can come together for a common local cause, but also that connectivity projects can function at a smaller scale and need not focus solely on charismatic megafauna to have an important impact on protecting biodiversity.

Finally, in an apt show of recognition for this hard-won achievement in corridor ecology, the Monkton Road Wildlife Crossing Project was recognized in 2017 by the Federal Highway Administration with an Environmental Excellence Award for "exemplary achievement [in] ecosystems, habitat, and wildlife."[29] This connectivity project was the product of an interdisciplinary working group comprised of the Lewis Creek Association; the Vermont Reptile and Amphibian Atlas; the town of Monkton, Vermont; VTrans; and other concerned publics. Since then, these amphibian crossing tunnels have continued to enable "thousands of salamanders and frogs to safely make their critical annual journey to and from their breeding pools."[30] The Monkton Wildlife Crossing demonstrates what is possible when grassroots efforts combine with evidence-based approaches to foster coexistence in an age of ever-increasing human development.

Notes

1. Slesar, "Movin' Lizards: 'Hey. You Movin' Lizards?" 18.
2. Slesar, "Movin' Lizards: 'Hey. You Movin' Lizards?" 18.
3. Slesar, "Movin' Lizards: 'Hey. You Movin' Lizards?" 18.
4. Banner Baird, "New Monkton."
5. Slesar, "Movin' Lizards: 'Hey. You Movin' Lizards?" 18; Lewis Creek Association, "Monkton Wildlife Crossing."
6. Slesar, "Movin' Lizards: 'Hey. You Movin' Lizards?" 18.
7. Vermont Fish & Wildlife Department, "Blue-Spotted Salamander."
8. VFWD, "Blue-Spotted."
9. VFWD, "Blue-Spotted."
10. VFWD, "Blue-Spotted."
11. VFWD, "Blue-Spotted."
12. Lewis Creek Association, "Monkton Wildlife Crossing."
13. For a broader and separate discussion of the State of Vermont's interagency working group formed in 2001 "to consider the public's mutual interests in effective wildlife conservation and transportation planning and development," see Austin, Slesar, and Hammond, "Strategic Wildlife Conservation and Transportation Planning: The Vermont Experience."

14 Banner Baird, "New Monkton."
15 Slesar, "Movin' Lizards: 'Hey. You Movin' Lizards?" 17.
16 Slesar, "Movin' Lizards: 'Hey. You Movin' Lizards?" 17.
17 Slesar, "Movin' Lizards: 'Hey. You Movin' Lizards?" 17.
18 Lewis Creek Association, "The Monkton Road Wildlife Crossing Project."
19 Kolb Noyes, "First Salamander Sighted."
20 Lewis Creek Association, "The Monkton Road Wildlife Crossing Project."
21 Slesar, "Movin' Lizards: 'Hey. You Movin' Lizards?" 22.
22 Slesar, "Movin' Lizards: 'Hey. You Movin' Lizards?" 22.
23 Slesar, "Movin' Lizards: 'Hey. You Movin' Lizards?" 22.
24 Slesar, "Movin' Lizards: 'Hey. You Movin' Lizards?" 22.
25 Slesar, "Movin' Lizards: 'Hey. You Movin' Lizards?" 22.
26 Conservation Planning, "Monkton Wildlife Crossing."
27 Slesar, "Movin' Lizards: 'Hey. You Movin' Lizards?" 22.
28 Slesar, "Movin' Lizards: 'Hey. You Movin' Lizards?" 22.
29 Slesar, "Movin' Lizards: 'Hey. You Movin' Lizards?" 22.
30 Lewis Creek Association, "Monkton Wildlife Crossing."

Chapter 7

THE RAILWAY FROM OXFORD TO LONDON MARYLEBONE

Transportation Upgrade Meets Compassion for Vulnerable Habitats

Thus far, we have focused largely on how roads and interstate systems can bisect habitats in ways that can prevent species from migrating or making their way to breeding sites; moreover, when roadways bisect habitats, species run the risk of losing their biodiversity through inbreeding. Within the field of road ecology, however, there has been less focus on the ways that railroads and railway systems can similarly disrupt ecosystems as a result of train tracks and tunnel systems being constructed in vulnerable habitats. Railway systems perhaps receive less attention because they are less consistently traveled than roads and highways.

Wildlife Implications of Railways Versus Roadways

At the same time, it is also inaccurate to make an easy comparison between cars and trains, as they and their environmental implications are far from the same, and each has different impacts on ecosystems and biodiversity. While train traffic may be less frequent than car traffic, trains travel much faster than cars. Additionally, interstates and railways are also structurally and physically different, "especially in the case of electrified railways, where overhead lines along the rail tracks can represent an additional source of impacts. All these differences are likely to affect wildlife responses to roads and railways."[1] And while trains are often viewed as a more climate-friendly and sustainable mode of transportation compared to vehicles, they too contribute to pollution, wildlife deaths, and habitat fragmentation. That said, statistically, railways are significantly safer than roadways, at least for people and cargo.

As described by Borda-de-Água et al., a 2013 report by the European Railway Agency notes that the "fatality risk in the period 2008–2010 measured as the number of fatalities per billion passenger-km is 0.156 to railway passengers and 4.450 for car occupants. In fact, following the same report, transport safety is only surpassed by the airline industry with 0.101 fatalities per billion passenger-km."[2] Railways are also more environmentally friendly than roads because a diesel-powered train is more energy efficient than its equivalent number of cars, and an electric-powered train is not a direct source of greenhouse gas emissions.[3] Railways generally also occupy less land than roads and interstates. However, while research has made clear the relative safety of railways in comparison with vehicles, for humans, there is a paucity of knowledge about wildlife mortality on railroad tracks in comparison with wildlife mortality on roadways. Such studies are few and far between and tend to focus on larger mammals, like train collisions with bears or moose.[4] In addition, "little is known about the consequences of vibration and noise on biodiversity living adjacent to the railway bed."[5] There is also a high rate of bird mortality from train collisions, based on "the assessment of bird carcasses found on railway tracks"; avian deaths due to train collisions most frequently involve raptors like hawks and owls and seabirds like ducks and gulls, but less so songbirds.[6] In fact, as Santos et al. describe: "In a study that compared mortality rates between roads and railways, it was found that railways had notably lower mortalities of songbirds, small mammals, and turtles when compared to those of roads"; this suggests that "diurnal and vagile species may be more efficient at avoiding trains than avoiding cars and trucks on busy two-way roads."[7]

Santos et al. make the apt point that it is also critical to learn more about the impact of high-speed trains, given their increased prevalence worldwide; they note that high-speed rails have different noise levels and fencing practices than traditional railways.[8] While more research is needed, both in the areas of traditional and high-speed railways' impacts on wildlife, more attention has been given in recent years to the ways that railways bisect critical habitats and impact wildlife biodiversity. That said, "compared to other transportation systems, such as roads, less is known about the impact of railways on wildlife, as well as its specificities," and railway ecology is considered an emerging area of research related to road ecology.[9] As a case in point regarding the increased awareness of the impact of railways on biodiversity, we may look to the recent work carried out by Chiltern Railways.

Chiltern Railways' Habitat Reconstruction Project

In October 2015, after upgrading some of its tracks, the company Chiltern Railways in the UK started to offer service between Oxford Parkway and London Marylebone. But in addition to the track upgrades and other repairs, the company added "purpose-built habitats" alongside the railway.[10] Not unlike the reconstruction of U.S. Highway 93 North in Montana, USA, this connectivity and restoration project had significant benefits for both people and wildlife; that is, the project not only provided a new commuting option for Oxford residents, but it also helped protect vulnerable natural habitats.

In particular, the Chiltern Railways habitat reconstruction project has benefited species including great crested newts, other reptiles, badgers, bats, and swallows. While the work has entailed some new landscape construction, the project has largely sought to minimize impacts on the surrounding natural habitat. The project has increased trackside habitat by 10,000 square meters, "including more than 11,000 trees, wildflower grasslands and hedgerows."[11] According to Andy Milne, senior program manager for Network Rail:

> As well as providing a great new option for Oxford commuters and significant economic benefits to the city, this scheme is trying to tread as lightly as possible on the natural environment. [...] When you're carrying out a major railway upgrade like this, cutting back lineside vegetation is unavoidable. But wherever we have had to fell a tree we replace it with an evergreen species more suitable for a modern railway corridor.[12]

In other words, while some biodiversity offsetting was required to complete the project, it was carried out with critical thought and without harm to plant or animal species.

Approaches to Habitat Conservation

New ponds have also been created to ensure the protection of great crested newts.[13] To this end, an independent ecological consulting firm first helped assess the "extensive ditch networks" along the railway, which "sit within a landscape well known for its widespread populations of great crested newt, a European Protected Species (EPS)."[14] In order to complete the project, the ecological consulting firm completed an assessment to ensure "that the conservation status of great crested newts would be unaffected by the development" and that measures would be taken to avoid harming them in any way.[15] After conducting the environmental assessment, they secured licenses to carry out the development work along the railway. This work involved installing newt fencing "across the development footprint, spanning over 21 km of line-side and adjoining fields. Amphibians are now being captured and excluded from the development footprint, and translocated to the previously created receptor areas."[16] As Tanith Cook, senior ecological consultant for the Chiltern Railways project notes:

"I think in some cases we have to accept that it is not always possible to replace 'like-for-like' habitat in the same location, especially if we all want better rail links and train services." [...] But, what is becoming more and more common is "biodiversity offsetting," which compensates for losses which cannot be avoided or mitigated within the rail corridor "by creating new habitats off site."[17]

As the ecology consulting firm involved with the Chiltern project describes, this undertaking required consistent and detailed communication with all stakeholders involved: "Throughout the life of the project, the complexities of ownership and responsibilities in connection with the project required extensive liaison with Natural England, tenants, landowners, clients and other stakeholders."[18] In order for the railway upgrades to proceed, the railway company had to ensure that the habitat of the great crested newts "would be unaffected by the development" and that no harm would come to this species.[19] Biodiversity offsetting is one area of conservation biology that can challenge some of the principles of compassionate conservation, depending on the project, scale, and level of displacement involved; however, as mentioned earlier in Chapter 3, this is not necessarily meant to suggest that such measures be ruled out entirely when it comes to connectivity projects. Rather, it means thinking critically about how to best ensure minimal disturbances to ecosystems, especially when some disturbance is inevitable. Thus, the current approach represents one possible path forward in achieving the goal of protecting habitats situated in railway environments.

Accounting for the Needs of Various Species

The railway habitat reconstruction project had to account for the needs of multiple species that reside in the area. Badgers who utilize the railway habitat were thus also accounted for in Chiltern's development plan. As a Chiltern representative describes, "10 artificial badger setts [dens] were built and are in use along the route. We have already seen young badger cubs being cared for by their mother next to one of the purpose-built setts."[20] Next, to protect swallow nesting habitats, "two replacement swallow nesting shelters have been constructed next to Oxford Parkway Station. At least four pairs of swallows have used this facility so far. Several broods have been spotted taking test flights near to the station building."[21] Notable too were efforts to "reinstate tree lines and linear flight paths used by commuting and roosting bats."[22] That is, the implementation of bat deterrent lighting at the Wolvercote Tunnel section of the railway also accounts for the impact of railways on *aero-corridors* for specific species.

According to Chiltern Railways, the bat deterrent lighting project "involved working closely with Natural England and their specialists to develop an innovative solution to manage the wellbeing of bats, by deterring them from flying around when a train is approaching and travelling through the tunnel."[23] The lighting project involved contracting the company Railway Electrical Services (RES) to carry out the work at Wolvercote Tunnel, which is registered in the UK as a "Bat Roosting Site."[24] The lighting that was installed prompts bats to hide when a train is nearing the tunnel.[25] To this end, RES installed specific lighting, power cabling, and power supplies and tested and certified the materials. The lighting project thus helped the bats maintain their linear flight paths and specific aero-corridors.

Compassionate Conservation and Railway Ecology

The Chiltern Railways habitat reconstruction project is unique in that it recognizes the ways that modes of urban transportation, such as railways, can have an impact on local habitats. As Cook comments, "Often, the area near a railway can be a relatively untouched habitat, with a range of habitats and high biodiversity. They can act as wildlife corridors."[26] The project is unique in its ability to account for issues related to urban transportation, corridor ecology, connectivity, and even the more burgeoning area of aero-corridors as they pertain to the flight paths of bats. The Chiltern railway project aligns with the principles of compassionate conservation, both in its goal of doing no harm to species and in valuing the lives of all species impacted by the work. While some of its approach involves biodiversity offsetting, such work was carried out with very specific planning and consideration for the lives and well-being of affected species, such as the great crested newt. In June 2017, the Bicester to Oxford Collaboration won the prestigious "Cross Industry Partnership Award" for their work on Chiltern Railways' Oxford to London service.[27] Ultimately, this railway connectivity project has sought to account for the needs of multiple species and has carried out a wildlife corridor reconstruction project that has helped further knowledge about the understudied area of railway ecology and wildlife corridors.

Notes

1 Borda-de-Água et al., *Railway Ecology*, viii.
2 Borda-de-Água et al., *Railway Ecology*, 4.
3 Borda-de-Água et al., *Railway Ecology*, 4.
4 Santos et al., "Current Knowledge on Wildlife Mortality in Railways," 6.
5 Borda-de-Água et al., *Railway Ecology*, 315.
6 Santos et al., "Current Knowledge on Wildlife Mortality in Railways," 16.

7   Santos et al., "Current Knowledge on Wildlife Mortality in Railways," 19.
8   Santos et al., "Current Knowledge on Wildlife Mortality in Railways," 20.
9   Borda-de-Água et al., *Railway Ecology*, 5.
10 Railway Technology, "Railways and Wildlife."
11 Network Rail, "Oxford to London Marylebone."
12 Network Rail, "Oxford to London Marylebone."
13 Railway Technology, "Railways and Wildlife."
14 BSG Ecology, "East West Rail Ecology Mitigation."
15 BSG Ecology, "East West Rail Ecology Mitigation."
16 BSG Ecology, "East West Rail Ecology Mitigation."
17 Railway Technology, "Railways and Wildlife."
18 BSG Ecology, "East West Rail Ecology Mitigation."
19 BSG Ecology, "East West Rail Ecology Mitigation."
20 Network Rail, "Oxford to London Marylebone."
21 Network Rail, "Oxford to London Marylebone."
22 Railway Technology, "Railways and Wildlife."
23 Chiltern Railways, "Game-changing New Railway Wins Leading Industry Award."
24 Railway Electrical Services, "Wolvercote Tunnel."
25 Railway Electrical Services, "Wolvercote Tunnel."
26 Railway Technology, "Railways and Wildlife."
27 Chiltern Railways, "Game-changing New Railway Wins Leading Industry Award."

Chapter 8

AERIAL CORRIDORS IN URBAN ENVIRONMENTS

Light Pollution and Migratory Birds

In the UK, the Chiltern Railways reconstruction project discussed earlier involved adding a bat deterrent lighting system at the Wolvercote Tunnel section of the railway to account for the well-being of resident bats. The lighting installation helped to deter the bats from flying into the tunnel or traveling through the tunnel when a train was approaching. The lighting project thus helped the bats maintain their linear flight paths and specific flyways. Ambient and artificial light can have important and specific impacts not only on bats but also on birds, who may be thrown off course or distracted by different kinds of light. The Chiltern Railways project considered with critical awareness the impact of lighting on resident bats and installed lighting specifically to prevent the bats from entering the Wolvercote Tunnel, where they might otherwise collide with oncoming trains. In that case, the lighting was a helpful deterrent that could help save their lives. In other contexts, however, artificial light pollution can have detrimental effects, especially for species of migrating birds.

In fact, the light emitted from skyscrapers and other buildings in urban environments can be especially hazardous for birds. According to the Cornell Lab of Ornithology, within the United States, an "estimated 600 million birds are killed every year from collisions with some of the country's largest skyscrapers."[1] In particular, Chicago, Illinois; Houston, Texas; and Dallas, Texas, are among the most dangerous cities in the U.S. for migrating birds. These cities are "uniquely positioned in the heart of North America's most trafficked aerial corridors."[2] As Hilty et al. note, there has been "some discussion around aero-corridors [...] although air space conservation is still in its infancy and terrestrial stepping-stone nonaerial corridors continue to be the focus of connectivity for most flying species."[3] Indeed, while this book has focused more on the issue of wildlife corridors in specific terrestrial landscapes, aerial corridors are an emerging area of focus for wildlife connectivity, and light pollution is one of the main disruptors for wildlife that must navigate them. This case considers the issue of light pollution and its impacts on aerial corridors for migrating birds and provides some possible suggestions for addressing this issue, with a focus on citizen education and "Lights Out" campaigns as possible starting points.

The Impacts of Light Pollution on Migrating Birds

To address the lack of quantified knowledge about light pollution levels on a global scale, Falchi et al. created a world atlas of artificial sky luminance, which they compiled with "light pollution propagation software using new high-resolution satellite data and new precision sky brightness measurements."[4] For the purposes of this discussion, light pollution may be defined as "the alteration of night natural lighting levels caused by anthropogenic sources of light," while "[n]atural lighting levels are governed by natural celestial sources, mainly the Moon, natural atmospheric emission (airglow), the stars and the Milky Way, and zodiacal light."[5] As the researchers describe, approximately "83% of the world's population and more than 99% of the U.S. and European populations live under light-polluted skies."[6] Moreover, the Milky Way is not visible to approximately one-third of the human population worldwide, "including 60% of Europeans and nearly 80% of North Americans."[7] Nighttime artificial light also contributes to increased bird mortality, as "almost one-half of the contiguous U.S. experiences substantial light-pollution during nighttime, including streetlights, safety lights and extensively lit buildings."[8]

Light pollution has even more pronounced implications during spring and fall migrations when billions of bird species migrate between North and South America. To do so, they rely on natural light from celestial sources—the moon, sun, and stars, in order to navigate, and artificial light pollution negatively impacts this process.[9] In North America, for instance, there are four main flyways, or "avian superhighways," which are generally understood to be the Atlantic, Mississippi, Central, and Pacific Flyways.[10] When migrating cross-country, birds seek out efficient pathways with ample places to rest. While certain bird species stay closer to coastal areas, others migrate along inland paths or follow larger rivers or mountain ranges.[11] Storms and wildfires can and do throw birds off course, but even so, they can generally navigate to their destination if they still have the strength and resources to get back on track.[12] For birds to successfully migrate, however, they do require navigable flyways or aerial corridors, and light pollution is highly detrimental to migrating birds. As Horton et al. describe in their study of migratory birds' exposure to artificial light: "The recent recognition of airspace as vital habitat–one that is subject to increasing modification by humans–highlights the fundamental need to understand how organisms cope with such alterations, [...] which pose numerous challenges to airborne organisms during periods of transit, including nocturnally migrating birds."[13] Fortunately, while understanding the impacts of airspace on bird conservation is a more nascent issue, it appears to be easily mitigated through at least one grassroots effort that involves people doing something as simple as conserving their lights.

Approximately half of the U.S. experiences photo-polluted nights; contributing light sources include "streetlights, safety lights, and extensively lit buildings," all of which negatively impact wildlife in a range of ways.[14] Structures such as towers, skyscrapers, and even lighthouses emit artificial light at night (ALAN) and attract nighttime-migrating birds, who can then become trapped in the illumination.[15] As the National Audubon Society describes, turning off bright lights, even for just a few minutes at a time, helps birds move on; this phenomenon was "discovered by the Cornell Lab of Ornithology and New York City Audubon during the annual 9/11 memorial in New York City. Hundreds of birds are caught in the memorial's beams every year but turning them off for just 20 to 30 minutes at a time greatly reduces the density of birds in the area."[16] For such night-migrating birds, "direct mortality as a result of collisions due to attraction to light [...] is the most obvious and direct effect of ALAN, but there are also more subtle effects, such as disrupted orientation [...] and changes in habitat selection."[17] Horton et al. recently conducted research on migrating birds' exposure to nighttime light pollution; their findings, especially related to urban areas, have great implications for shaping approaches to "conservation actions to identify locations where ALAN-reducing programs may be most effective."[18]

Conservation Approaches Aimed at Reducing Light Pollution

Horton et al. found that migrating birds are most at risk to ALAN in the fall when migration numbers are highest, but "shifts in migratory routes between spring and fall migration also affect the numbers of birds exposed to higher light levels."[19] In the fall, these more easterly routes tend to take birds "over more heavily photo-polluted areas than do spring routes, leading to even higher numbers of birds—and many young birds—exposed to ALAN in fall."[20] Because exposure to light pollution is greatest overall in the fall, conservation and education efforts, including Lights Out campaigns, would likely be most effective during this time. On the other hand, Horton et al. found that "the patterns in Los Angeles and other cities in California are the opposite of most East Coast cities, with higher exposure during spring migration."[21] In addition, migration time frames can span over six months, "with hundreds of millions of individual migrants aloft on a given night," but oftentimes birds' migration "occurs in sporadic waves" or at very specific sites "during just a few peak nights."[22] One potential advantage of the ability to predict the time and place of migrations, specifically, is that it becomes possible to educate the public about the benefits of reducing light pollution during these times. Thus, the researchers note the importance of directing conservation and education efforts to times and locations where they will be most effective and to "where the highest numbers of birds are exposed to the highest amounts of ALAN."[23] As mentioned earlier, metropolitan Chicago, Houston, and Dallas are all at the top of the light pollution exposure risk ranking.

Compassionate Conservation and Lights Out Campaigns

The ability to predict when and where the greatest number of migratory birds will pass through a metro area allows cities to make concerted efforts to conserve lights and educate the public at key points in time when such efforts could be most effective and meaningful. To this end, Houston Audubon has begun using "migration forecasts" from the Cornell Ornithology Lab's *BirdCast* program to implement "lights out" warnings on nights when large numbers of migratory birds are expected to fly over the city.[24]

Chicago has also implemented a successful Lights Out program that has saved the lives of countless migratory birds who pass through the city during their migration. That is, "twice a year, around five million birds, representing about 250 species, fly through Chicago."[25] The lights from Chicago skyscrapers can confuse birds like the song sparrow, which is a common bird to migrate through the city at night. While lighting is considered an important feature of a building's architecture, prominent buildings like the John Hancock Building and the Sears Tower turn off their lights when migrating birds are known to be making their way through the city.[26] Chicago was an early adopter of a Lights Out program, which

> encourages building managers to dim or turn off decorative lighting late at night and to minimize the use of bright interior lights during migration season. They also encourage high-rise residents to draw their shades or dim interior lights late in the evening. The program was started in the fall of 2000 and has won the support of almost all of the major skyscrapers in Chicago.[27]

There is anecdotal evidence that the program helps. A representative from the Bird Collision Monitor and Rescue Project described one night when there was a tragic mix up and the lights did not go out on schedule. We had a heavy night of migration and there were birds everywhere with few survivors. The next night, the lights were out and the drop in fatalities and injuries was amazing—we guess at an 80 percent drop. We could still hear the birds flying, but they made it safely over the buildings that night.[28]

Another study conducted by the Field Museum in Chicago "found that turning off the lights at a downtown high-rise reduced migratory bird deaths there by 80 percent."[29]

The National Audubon Society originally spearheaded the Lights Out initiative as a national effort to reduce light pollution, especially for migratory birds, and the effort has since been adopted by many cities in North America. Some of their recommendations, geared toward the general public, include:

Turn off exterior decorative lighting; Extinguish pot and flood-lights; Substitute strobe lighting wherever possible; Reduce atrium lighting wherever possible; Turn off interior lighting especially on higher stories; Substitute task and area lighting for workers staying late or pull window coverings; Down-shield exterior lighting to eliminate horizontal glare and all light directed upward; Install automatic motion sensors and controls wherever possible; When converting to new lighting assess quality and quantity of light needed, avoiding over-lighting with newer, brighter technology.[30]

Lights Out campaigns are arguably a grassroots rendering of compassionate conservation at work. Cities, individuals, building tenants, and building managers can reduce light pollution in an effort to show that every bird species matters when it comes to successfully navigating through urban areas during peak migration seasons. The lead researcher in the study of migratory birds' exposure to artificial light, Kyle Horton, puts it this way:

If you don't need lights on, turn them off. [...] It's a large-scale issue, but acting even at the very local level to reduce lighting can make a difference. While we're hopeful that major reductions in light pollution at the city level are on the horizon, we're excited that even small-scale actions can make a big difference.[31]

Reducing light pollution also has benefits that extend beyond helping birds navigate through cities—such initiatives also save a large amount of energy. In 2006, a municipal building that participated in Toronto's Lights Out program "reported cost savings of more than $200,000," while "turning off a single 100-watt bulb from dusk until dawn saves an average of 417 kwH of electricity, or $46 a year in bulbs and electricity costs."[32]

When it comes to the more nascent issue of airspace conservation as it pertains to aerial corridors, citizen education, and public outreach campaigns have emerged as viable paths forward in the effort to reduce light pollution. Moreover, Lights Out campaigns not only represent grassroots conservation efforts aimed at helping birds migrate through urban areas, but they also provide educational opportunities to promote knowledge, understanding, and empathy for vulnerable migratory bird species. Finally, such efforts at public outreach and education help illustrate that it is possible to help foster connectivity even at the individual level by making relatively minor adjustments to our ways of living at a time when even seemingly small human actions constitute major threats to our nonhuman kin.

Notes

1. Dapcevitch, "American Skyscrapers."
2. Dapcevitch, "American Skyscrapers."
3. Hilty et al., *Corridor Ecology*, xiii.
4. Falchi et al., "The New World Atlas of Artificial Night Sky Brightness," 1.
5. Falchi et al., "The New World Atlas of Artificial Night Sky Brightness," 1.
6. Falchi et al., "The New World Atlas of Artificial Night Sky Brightness," 4.
7. Falchi et al., "The New World Atlas of Artificial Night Sky Brightness," 1.
8. Dapcevitch, "American Skyscrapers."
9. Dapcevitch, "American Skyscrapers."
10. Fritts, "Avian Superhighways."
11. Fritts, "Avian Superhighways."
12. Fritts, "Avian Superhighways."
13. Horton et al., "Bright Lights in the Big Cities," 209.
14. Horton et al., "Bright Lights in the Big Cities," 209.
15. Horton et al., "Bright Lights in the Big Cities," 209.
16. Audubon, "Lights Out."
17. Horton et al., "Bright Lights in the Big Cities," 209.
18. Horton et al., "Bright Lights in the Big Cities," 212.
19. Horton et al., "Bright Lights in the Big Cities," 212.
20. Horton et al., "Bright Lights in the Big Cities," 212.
21. Horton et al., "Bright Lights in the Big Cities," 212.
22. Horton et al., "Bright Lights in the Big Cities," 213.
23. Horton et al., "Bright Lights in the Big Cities," 213.
24. The Cornell Lab of Ornithology, "Study Names Top Cities."
25. Kousky, "Chicago Skyscrapers."
26. Kousky, "Chicago Skyscrapers."
27. Kousky, "Chicago Skyscrapers."
28. Kousky, "Chicago Skyscrapers."

29 Golden Gate Bird Alliance, "Lights Out for Birds."
30 Audubon, "Lights Out."
31 The Cornell Lab of Ornithology, "Study Names Top Cities Emitting Light that Endangers Migratory Birds."
32 Golden Gate Bird Alliance, "Lights Out for Birds."

Chapter 9

THE PAPAHĀNAUMOKUĀKEA MARINE NATIONAL MONUMENT

Traditional Ecological Knowledge and Marine Protected Areas

The Papahānaumokuākea Marine National Monument (PMNM) is located in the northwestern part of the Hawaiian Archipelago and is one of the world's largest marine protected areas. The Northwestern Hawaiian Islands (NWHI) consist of a group of small, remote islands and atolls northwest of the Kauai and Niihau islands and include one of the most pristine coral reef ecosystems in the world. This ecosystem supports a large number of apex predators and other endemic species and is a critical habitat for many threatened and endangered species. In June 2006, almost 140,000 square miles of this marine environment were designated as the Northwestern Hawaiian Islands MNM; one year later, the area was renamed Papahānaumokuākea.[1]

The PMNM also encompasses the Midway Atoll, which has been in the media in recent years due to the vast quantities of marine plastic debris that litter its shores and threaten the Laysan Albatross that nests in the NWHI region (Figure 9.1). In addition to the more well-known Laysan Albatross, however, the national monument is also a transit corridor that is used by approximately 14 million seabirds of different species for breeding, foraging, and stopping over during migrations.[2] The Northwestern Hawaiian Islands also provide an important stopover habitat for shorebirds as they migrate through the central Pacific and are home to the Laysan Finch, Nihoa Finch, and Nihoa Millerbird, which are all endangered and found only on "one or a few islands, putting their populations at risk from predators, storms, and other catastrophic events."[3] Finally, "at least six species of terrestrial plants found only in the region are listed under the U.S. Endangered Species Act, some so rare that because of the difficulty of surveying these remote islands, they have not been documented for many years."[4]

Figure 9.1 Laysan Albatross colony on Papahānaumokuākea Marine National Monument, Midway Island, Midway Atoll, Hawaiian Islands. Credit: Gerald Corsi.

The Creation of the Papahānaumokuākea Marine National Monument (PMNM)

At the time that PMNM was created in 2006, it already "covered 140,000 square miles of ocean around the uninhabited northwestern islands of Hawaii."[5] Then, in 2016, former president Barack Obama expanded the boundaries of the national monument, which "more than quadrupled Papahānaumokuākea's size, to 582,578 square miles, an area larger than all the national parks combined."[6]

The expansion of the PMNM in the Northwest Hawaiian Islands established the world's largest protected marine reserve. Expanding the MPA helps ensure further protection of these interconnected marine ecosystems, which are "comprised of deepwater and shallow habitats, coral reefs, low-lying atolls and islands"; moreover, the monument is estimated to support about 7,000 species, "including endemic corals, fish, marine reptiles and mammals," in addition to bird species.[7] It is important to note that "coral atolls may appear as separated islands but are connected across vast distances to form functional marine ecological networks."[8]

The boundary expansion of Papahānaumokuākea National Monument was politically and culturally significant for several reasons. The expansion leveraged the 100-year-old Antiquities Act, which was first signed by former president Theodore Roosevelt in 1906 and was created to "preserve and protect critical natural, historical, and scientific resources on Federal lands for future generations."⁹ Just a few years later, in 1909, President Roosevelt also created the Hawaiian Islands Bird Reservation, in part because of the taking of seabirds on these islands. Two of those national wildlife refuges, the Hawaiian Islands National Wildlife Refuge and the Midway Atoll National Wildlife Refuge, are within the PMNM area.

With the expansion of the geographic boundary also came an expansion of the "no-take" area in which mineral extraction and commercial fishing were prohibited. This was viewed by many as a controversial move, as it was perceived by some as a threat to Hawaii's fishing culture. As Edwin Ebisui Jr., chair of the Western Pacific Regional Fishery Management Council, put it: "To the native Hawaiian, access to marine resources is very, very important and always will be [...] I don't see how quadrupling the size of the prohibited fishing area in any way furthers their cultural interests."[10] While deep-sea mining and commercial fishing are prohibited within the boundaries of the protected area, "recreational fishing and subsistence fishing by native Hawaiians" can be carried out by permit, as well as scientific research.[11]

Management of the PMNM

The monument is co-managed by four main co-trustees and overseen by other national, regional, and local partners. According to the 2021 White House memo "Indigenous Traditional Ecological Knowledge and Federal Decision Making," the four co-trustees are the National Oceanic and Atmospheric Administration, U.S. Fish and Wildlife Service, the State of Hawai'i Office of Hawaiian Affairs, and the Hawai'i Department of Land and Natural Resources; additionally, "Native Hawaiians have consistently led the development and governance of the monument."[12] These groups are "committed to preserving the ecological integrity of the Monument and perpetuation of the NWHI ecosystems, Native Hawaiian culture and other historic resources" and work together to carry out the PMNM's mission.[13] In addition, a Memorandum of Agreement (MOA) was signed in 2006 by the Secretary of Commerce, Secretary of the Interior, and the Governor of Hawai'i, which "provides for coordinated administration of all the Federal and State lands and waters within the boundaries of the Monument. The MOA established the institutional arrangements for managing the Monument, including representation of Native Hawaiian interests by the Office of Hawaiian Affairs on the Monument Management Board."[14] In 2008, the co-trustees also created the PMNM Management Plan to help manage the monument's ecological, cultural, and historic resources. The PMNM Management Plan "provides a framework" for understanding the impacts of climate change on this marine environment and provides a foundation for developing "adaptation strategies to these impacts."[15]

The co-management structure of the PMNM also aims to give "indigenous people a greater say over the stewardship of the lands and waters they have inhabited for generations."[16] With the boundary expansion of the PMNM, many lawmakers in Hawaii felt that the marine protected area (MPA) could serve as a model of both cultural and ecological sustainability for the world's oceans.[17] The boundary expansion was significant, then, not only for its positive environmental impacts but also because the Northwestern Hawaiian Islands carry much cultural significance and are viewed by native Hawaiians as a sacred place.[18] In fact, Papahānaumokuākea itself is considered a very sacred area.

Papahānaumokuākea: A Sacred Place with a Sacred Name

According to Hawaiian tradition, "Native Hawaiian culture is living—it is the expression of a people that continues to evolve in great part through the perpetuation of a rich, oral tradition" carried out through chants and songs that tell the stories of Hawaiian history, natural resources, and spiritual knowledge.[19] In fact, a Hawaiian chant called "the Kumulipo" tells the story of how all life evolved from Papahānaumokuākea, starting with coral; therefore, "the genealogy of Papahānaumokuākea tells the story of Native Hawaiians' ancestral connection with the gods who created those coral polyps, the Northwestern Hawaiian Islands or Kūpuna (respected elders) Islands, and everything else in the archipelago, including Native Hawaiians."[20] Papahānaumokuākea is considered a sacred place and has many interconnected, culturally significant sites throughout its ten atolls and islands.[21]

Naming Papahānaumokuākea

The name Papahānaumokuākea, pronounced "Pa-pa-hah-now-mo-koo-ah-keh-ah," is rooted in an ancient Hawaiian tradition related to the "genealogy and formation of the Hawaiian Islands, and a deep honoring of the dualisms of life." To this end, "Papahānaumoku is a mother figure personified by the earth and Wākea is a father figure personified in the expansive sky; the two are honored and highly recognized ancestors of Native Hawaiian people. Their union resulted in the creation, or birthing, of the entire Hawaiian archipelago"; therefore, naming the monument is seen as a way to honor these names and preserve Hawaii's cultural heritage.[22]

The naming of the Northwestern Hawaiian Islands Marine National Monument began in 2006, when the sanctuary itself was designated, at this time, the Northwestern Hawaiian Islands Native Hawaiian Cultural Working Group selected two distinguished members of the Hawaiian community: Uncle Buzzy Agard and Aunty Pua Kanahele, to contribute their ideas for names. The Northwestern Hawaiian Islands Native Hawaiian Cultural Working Group itself is made up of members who have long-term involvement in and knowledge of the region and have "varied relevant backgrounds, and include academic scholars, teachers, cultural practitioners, community activists, and resource managers that have experience in working directly with issues concerning the Northwestern Hawaiian Islands."[23] As Bekoff notes, compassionate conservation "involves people with wide-ranging interests—both academics and nonacademics—working hard together to balance the well-being of people, individual animals, and the health and integrity of landscapes and ecosystems. Academics, advocacy, and activism go hand in hand."[24] After the initial ideas for names were suggested, the Cultural Working Group selected the name they felt would be most appropriate for "the Northwestern Hawaiian Islands managing entity and region," and on January 4, 2007, the group chose Papahānaumokuākea.[25]

Accounting for Climate Change

Despite the challenges of accounting for the unknown, future impacts of climate change on the region, the PMNM Management Plan does provide "a framework for integrating strategies that focus on climate change," which includes research, monitoring, education, and outreach. More specifically, the management plan recommends the following research activities to analyze the impacts of climate change on the NWHI:

> Determine the effect of sea level rise on nesting sites of protected species (e.g., Hawaiian monk seal, green sea turtle); Identify specific habitats, communities, and populations that will be affected by climate change; Identify habitat changes that will result from sea level rise; Map areas that will be most affected by extreme weather events; and Distinguish human-made impacts from natural variability of the biophysical environment.[26]

Coral reefs in particular are the subject of long-term research, and such research and monitoring programs are continuously conducted within the NWHI by the Coral Reef Ecosystem Division of the NOAA Pacific Islands Science Center; the Coral Reef Ecosystem Integrated Observing System, which focuses on mapping and visual observation; and Coral Reef Watch, which detects bleaching "hotspots."[27] In June 2016, following the request of the management team, a "formal assessment of climate-related vulnerabilities" was conducted, and the results of that research "noted with a high degree of confidence that the most vulnerable resources within the monument to climate change are coral reefs, low-lying atolls, and beaches"; these areas include habitats for "endangered birds, monk seals, sea turtles, and archeological sites and artifacts."[28]

Conclusion

The PMNM has been precedent-setting in scope and scale, and in terms of the range of species and ecosystems it has helped to protect. Its management plan recommends strategies to help account for climate change, and its management approach takes into account native Hawaiian cultures and traditions. Media accounts of the PMNM have focused largely on the significance of the recent boundary expansion as emblematic in working against climate change and protecting oceans and marine life. As Sarah Chasis, director of the oceans program at the Natural Resources Defense Council, put it: "This act—to build resilience in our oceans, and sustain the diversity and productivity of sea life—could usher in a new century of conservation for our most special, and fragile, ocean areas."[29]

The expanded MPA will certainly help protect marine species like the albatross from the threats of longline fishing; however, the move is not a panacea. That is, on the one hand, there is no single fix that will undo the anthropogenic effects of climate change; on the other hand, this doesn't mean we ought not to try. President Obama's expansion of the MPA helped protect the region from certain threats like longline fishing and deep-sea mining, and it has helped garner additional attention around the issue of ocean plastics and vulnerable species in the region.

The boundary expansion, though, is but one piece of a much larger puzzle. For, while a larger MPA should help protect albatrosses from the threat of longline fishing, for instance, it still does not prevent plastics from washing ashore on Midway Island, or, as one National Geographic article noted: "Being inside a national monument doesn't protect Laysan Island from trash washing ashore, especially on its windward side."[30] That is, ocean currents and the plastics circulating within them are indifferent to the nuances of geographic boundary-making for the purposes of environmental policy and protection. Even so, the establishment of this marine protected area helps chart a productive path forward in conservation management grounded in evidence-based decision-making, compassion, and traditional ecological knowledge.

Most recently, in 2021, the co-trustees of the PMNM released a historic guidance document titled "Mai Ka Pō Mai," intended to help federal and state governments integrate Native Hawaiian culture into all areas of management. The document was developed through regular meetings with the Native Hawaiian community over the past decade and will provide the foundation for the future update of the PMNM's Management Plan.[31]

Compassionate conservation advocates that we question with critical awareness the role of human intervention in efforts to control, manage, or recuperate the natural world. Approaches to conservation management that eschew speciesism, take a non-hierarchical approach to protecting species and places, and incorporate TEK, such as the PMNM has done, are implicitly aligned with a compassionate conservation approach. The PMNM has also shown that it is possible to account for climate change in the management plan of a marine protected area; however, given the unpredictability of climate change and natural disasters, it is difficult to predict what aspects of such a plan will be most relevant and to what extent such plans may be implemented—that does not mean we ought not to try, however. This case has highlighted the importance of accounting for marine systems when thinking about corridor ecology, and it illustrates the value of traditional ecological knowledge in conservation planning.

Notes

1 Pacific Birds, "Birds of Papahānaumokuākea."
2 Pacific Birds, "Birds of Papahānaumokuākea."
3 Pacific Birds, "Birds of Papahānaumokuākea."
4 National Ocean Service, "Papahānaumokuākea Marine National Monument."
5 Barnett, "Hawaii is Now Home."
6 Barnett, "Hawaii is Now Home."
7 Pacific Birds, "Birds of Papahānaumokuākea."
8 Hilty et al., *Guidelines*, 28.
9 U.S. Department of the Interior, "Antiquities Act."
10 Barnett, "Hawaii is Now Home."
11 Hirschfeld Davis, "Obama to Create World's Largest Marine Reserve."
12 Executive Office of the President, "Memorandum."
13 State of Hawai'i: Division of Aquatic Resources, "Papahānaumokuākea."
14 National Ocean Service, "Management."
15 Kershner, "Incorporating Climate Change Adaptation."
16 Hirschfeld Davis, "Obama to Create World's Largest Marine Reserve."
17 Barnett, "Hawaii is Now Home."
18 Hirschfeld Davis, "Obama to Create World's Largest Marine Reserve."
19 National Ocean Service, "Papahānaumokuākea: A Sacred Name, A Sacred Place."
20 National Ocean Service, "Papahānaumokuākea: A Sacred Name, A Sacred Place."
21 National Ocean Service, "Papahānaumokuākea: A Sacred Name, A Sacred Place."
22 National Ocean Service, "Papahānaumokuākea: A Sacred Name, A Sacred Place."
23 National Ocean Service, "Papahānaumokuākea: A Sacred Name, A Sacred Place."
24 Bekoff, *Rewilding*, 60.

25　National Ocean Service, "Papahānaumokuākea: A Sacred Name, A Sacred Place."
26　Kershner, "Incorporating Climate Change Adaptation."
27　Kershner, "Incorporating Climate Change Adaptation."
28　Kershner, "Incorporating Climate Change Adaptation."
29　Hirschfeld Davis, "Obama to Create World's Largest Marine Reserve."
30　Barnett, "Hawaii is Now Home."
31　Papahānaumokuākea Marine National Monument, "Integrating Native Hawaiian Culture into Management of Papahānaumokuākea."

Chapter 10

LOOKING AHEAD

New Perspectives and Best Practices Related to Wildlife Corridors

This chapter brings together the concepts, philosophies, and case examples throughout the book to consider a working set of best practices and new theoretical directions for wildlife corridor and connectivity projects. As we consider the theories and illustrative examples that this book has outlined, it is clear that connectivity projects are highly context-dependent in terms of the landscapes, communities, and stakeholders involved. Even so, as I hope the explorations in this book have shown, it is possible to identify a productive path forward for corridor projects—namely, one that privileges coexistence and engages more directly with theories and practices that involve compassionate conservation, empathy, and traditional ecological knowledge (TEK).

In the Introduction, I acknowledged that wildlife corridors can be defined in different ways depending on the context, may occur naturally or be human-made, and can happen at different levels of scale. On the whole, wildlife corridors "provide continuous habitat for species to move on their own, [and] are a reasonable and effective means for ensuring connectivity in the landscape."[1] I acknowledged that this book does not necessarily advocate for perfect, tidy corridor projects; rather, by being mindful of approaches informed by compassionate conservation, empathy, and TEK, we can conceptualize and implement corridor projects that engage more holistically with landscapes, wildlife, and people and better foster coexistence and connectivity.

An approach informed by compassionate conservation understands that the lives of all beings, human and nonhuman, matter in decisions about wildlife conservation and policy; moreover, when it comes to humans making decisions about the lives of nonhuman animals, a compassionate conservation approach would advocate for an approach to corridors that values all nonhuman animals—not just more well-known megafauna or keystone species; finally, it advocates that, whenever feasible, corridor and conservation projects employ solutions that proceed with as minimal disturbance to the natural landscape as possible. Ideas about empathy and TEK are implicitly aligned with such an approach and understand living beings, including humans, as needing to coexist with one another and with their environment. Entangled empathy similarly focuses on another's experiential well-being in the world and tends to be action-oriented and interested in what seems to be the best, or most compassionate choice, in helping to pursue or participate in another's well-being. In the Introduction and throughout the book, I argue that if designed with the principles of compassion and empathy in mind and with the voices of multiple stakeholders represented, including Indigenous perspectives, wildlife corridors have the potential to perform conservation practice grounded in an ethic of care and empathy, as well as to reshape how we understand what counts as "natural" places and our ideas about "who belongs where."

Chapter 2 described some of the key concepts underpinning wildlife corridors in an effort to lay the groundwork for a further understanding of the theoretical concepts and illustrative cases that can inform new ideas about wildlife corridor planning and design. Namely, the chapter addressed the concepts of connectivity and corridor habitats more broadly, landscape fragmentation, structural and functional connectivity, gene flow, scale, types of corridors, and the idea of green infrastructure. It held a closer lens to the Yellowstone to Yukon (Y2Y) corridor project, not only as an initial example of how TEK can help shape the success of connectivity projects but also as an example that illustrates that clear management and communication are possible, even with larger-scale projects that span countries and ecosystems.

Chapter 3 began with a short example of the Parleys Summit wildlife crossing in Utah, in an effort to illuminate some emergent perspectives and questions related to the planning and design of corridor projects, as well as to help illustrate our different rationales for valuing wildlife corridor projects in the first place. In preparation for Chapter 4, I began to consider our rationales for how we value our relationships with wildlife and the extent to which our rationales may be more focused on the needs and interests of humans or nonhuman animals; for instance, I questioned whether a human-centric rationale must always be present in order to also protect wildlife, and if so, whether that is necessarily a negative outcome. Ultimately, I suggest that whenever possible, wildlife corridors should attempt to account for the best interests of all stakeholders, which often encompass both people and wildlife; this entails considering landscape elements as well as socioeconomic and political contexts that shape corridor design and implementation, along with the needs of not only flagship species but also all wildlife who will make use of the corridor, when feasible. It also describes the need to minimize habitat disturbance as much as possible but acknowledges that, in some instances, it is not always possible to completely avoid disturbances to species in habitat reconstruction projects.

Chapter 3 winds down by providing some general design guidelines for corridors, which entail the following: focus on connecting patches that were previously connected; avoid connecting artificial patches with less-disturbed, higher-quality habitat; seek out and preserve currently existing natural corridors; design corridors along altitudinal and latitudinal gradients to achieve the greatest possible biodiversity and help mitigate climate impacts; and avoid long expanses of landscape without nodes or endpoints like habitat patches, and incorporate redundant connections through the use of alternate pathways.[2] Clearly, these guidelines are just that—guidelines, and they may not be relevant or feasible for all corridor projects. Finally, the chapter suggests focusing on communication across groups and audiences in an effort to understand the various priorities that inform conservation goals. Issues related to land rights, for instance, should be approached with care and inclusive communication, as the Y2Y project has demonstrated. Such issues also illustrate the need to incorporate TEK into corridor design, which leads to some of the emerging theories and perspectives discussed in Chapter 4.

Chapter 4 held a closer lens to ideas about compassionate conservation, empathy, and TEK, which have been discussed in previous chapters, and informed each of the subsequent illustrative cases in different ways. The chapter expanded on ideas about compassionate conservation by also integrating the concept of rewilding, which, when applied to wildlife corridors, considers that "(1) healthy ecosystems need large carnivores, (2) large carnivores need big, wild roadless areas, and (3) most roadless areas are small and thus need to be linked."[3] Here, Bekoff understands rewilding as "a large-scale process involving projects of different sizes that may focus on carnivores but ultimately include a panoply of wildlife."[4] The chapter also explored the ways that empathy is aligned with both compassionate conservation and TEK; it addressed critiques of compassionate conservation and considered how an openness to learning about the specific contexts of nonhuman animals and an interest in learning about the processes that shape the lives of vulnerable nonhuman species and their habitats can help work against bias or projection. I suggest that wildlife corridor design and planning constitute evidence-based, decision-making frameworks for corridor ecology that encourage a potential middle ground when it comes to conservation projects that aim to restore connectivity and preserve biodiversity. Moreover, as some of the illustrative cases in this book show, many of the stakeholders who speak on behalf of corridor projects implicitly and explicitly express a combination of values related to the need for wildlife corridors, along with the need for an action-based approach. TEK can inform such approaches through its emphasis on how humans relate to, coexist with, and advocate for the lives of nonhuman animals and through its concern with the relationships of all living beings, including humans, with one another and with their environment.

Next, the illustrative case chapters in the book helped further contextualize the ideas above in different ways. They explored some specific examples of how wildlife corridors may function in sync with these ideas and philosophies and how such work can benefit landscapes, wildlife, and people. In Chapter 5, for instance, the connectivity project at the Flathead Indian Reservation illustrated the value of including the voices of Indigenous peoples and TEK in the management of this conservation area and the I-93 reconstruction project, which in turn helps support coexistence and compassionate conservation. In Chapter 6, the Monkton Wildlife Crossing Tunnels show the importance of smaller-scale corridor projects in their ability to help preserve biodiversity—this corridor project shows how concerned citizens can work at a more grassroots level to implement corridor projects and that such projects need not focus solely on charismatic megafauna to have an important impact on protecting biodiversity. In Chapter 7, the Chiltern Railways upgrade project sheds light on the less-studied area of railway ecology as applied to wildlife corridors. This habitat reconstruction project benefited multiple species, including great crested newts, other reptiles, badgers, bats, and swallows, and did so in ways that involved critical awareness of the lives and well-being of these species. While this book has focused more on the issue of wildlife corridors in specific terrestrial landscapes, Chapter 8 explored the nascent area of aerial corridors relative to wildlife connectivity—namely, the impacts of light pollution on migratory birds. It provided some possible suggestions for mitigating this issue and focused on citizen education and "Lights Out" campaigns as possible starting points. Finally, Chapter 9 explored the Papahānaumokuākea Marine National Monument (PMNM)—a precedent-setting project in terms of scope, scale, and the marine ecosystems involved. The project has not only shown that it is possible to account for climate change in the management plan of a marine protected area (MPA), but it has also illuminated the importance of accounting for marine systems when thinking about corridor ecology. Moreover, the PMNM has been precedent-setting in its ability to illustrate the value of TEK in marine conservation planning—an area that continues to emerge as critical for subsequent generations of conservation and corridor management projects.

Taken together, the case examples provided throughout this book help contextualize many of these practices and show that it is indeed possible to come together and undertake corridor projects that account for multiple perspectives, and account for the best interests of the affected habitats and the beings, nonhuman and human, who inhabit them.

Looking Ahead: The Proposed Chumash Heritage National Marine Sanctuary

At the time of this writing, another project of marine conservation is taking shape off the coast of California. The Northern Chumash Tribal Council is currently requesting federal protection for 7,000 square miles of territory along 156 miles of California's central coast. If their request is approved, the Chumash Heritage National Marine Sanctuary (CHNMS) would be the first Tribal-nominated national marine sanctuary designation in the United States.[5] The push for a national marine sanctuary in Central California is not a recent undertaking; community members and elected leaders have been advocating for such a national marine sanctuary in Central California for decades. In 2013, the Northern Chumash Tribal Council launched their campaign for a formal sanctuary designation.[6] In 2015, the National Oceanic and Atmospheric Administration (NOAA) accepted their nomination, and the sanctuary was placed on a list for future consideration.[7]

As the CHNMS moves through the review process, federal agencies will assess NOAA's "draft regulations detailing the proposed terms of the sanctuary."[8] A public comment period is open through fall 2023; if the sanctuary is approved, NOAA would then review and incorporate suggestions from the public and oversee the creation of the sanctuary, and California's governor along with Congress would make their own recommendations. The CHNMS would then benefit from "increased government resources for ecological research, public education and outreach, and operating a visitors' center to teach the public about the importance of conserving ocean waters," according to Paul Michel, regional policy coordinator for NOAA sanctuaries' West Coast region.[9] As Michel also noted, NOAA is interested in new ways of incorporating Chumash culture in the creation of the proposed marine sanctuary, which may include placing Chumash translations on sanctuary signage and incorporating Chumash history in educational programming.[10] In other words, the story of the sanctuary and its significance would be conveyed "through the eyes of the stewards of this coast for 10,000 years" as Michel described.[11]

The proposed CHNMS would further highlight the importance of accounting for Indigenous perspectives and cultural values in ocean conservation. The protections provided by the CHNMS could also help alleviate some of the threats posed to marine life by human development, including offshore oil development and ocean noise caused by acoustic testing.

Consistent with the 2021 memo issued by the White House Office of Science and Technology Policy and the White House Council on Environmental Quality titled "Indigenous Traditional Ecological Knowledge and Federal Decision Making," the proposed sanctuary would "help to address the global biodiversity crisis and accelerate nature-based solutions through inclusive partnerships and collaboration among federal, state and local governments, and California Native American Tribes."[12] In doing so, "Chumash tribes would gain a unique leadership role over an expansive marine sanctuary, including the ability to block unwanted commercial development on the land and water within its bounds."[13]

The establishment of MPAs can help preserve ocean biodiversity because "MPA networks encompass multiple instances of habitats and species" that constitute similar communities "and are linked by larval dispersal."[14] Because marine species "associate with particular habitats, MPA networks designed to protect biodiversity and facilitate population and community connectivity require that a variety of coastal ecosystems be represented throughout the network."[15] The proposed CHNMS would leverage TEK and an implicitly compassionate conservation approach to protect biodiversity within its designated waters. The proposed marine sanctuary would protect biodiversity in an area of the Pacific Coast where "temperate waters from the north meet warmer southern currents, and seasonal upwelling of nutrient-rich waters along the California Current fuel the food web, supporting a rich marine ecosystem."[16] This region of the ocean is also known to be a

> biological hotspot for birds, marine mammals, sea turtles, fishes, other marine organisms, and algae, like kelp. More specifically, the area includes thriving ecological habitats such as kelp forests, rocky reefs, and sandy beaches, as well as unique and important offshore geologic features like the Rodriguez Seamount, Santa Lucia Bank, and Arguello Canyon, all home to unique and rare deep-sea corals and sponges

as well as feeding areas for 13 species of whales and dolphins.[17] A large portion of the California sea otter population also inhabits this region and feeds on its kelp forests, and the area is also home to large numbers of pinnipeds including a significant population of harbor seals.[18]

Of additional relevance to the success of this proposed sanctuary is the fact that it sits between already existing MPAs, which will further enhance ecological connectivity and protected corridors for fish and wildlife.[19] More specifically, the proposed area of protection sits between the Channel Islands National Marine Sanctuary and the Monterey Bay National Marine Sanctuary and out to the western slope of the Santa Lucia Bank; the proposed sanctuary blends cultural knowledge with biodiversity needs and "warrants protection under the National Marine Sanctuary Program for the purpose of embracing the Chumash concept of 'thrivability' wherein a deep understanding of this unique and precious marine environment is embodied within its local human inhabitants."[20] As such, the CHNMS would also be inclusive of TEK and incorporate such knowledge into its management and decision-making processes by drawing on deep knowledge of Chumash culture and regional ocean biodiversity. That is, Chumash culture has a long and rich history "dedicated to the nurturing of relationships to Nature and the Ocean in the deepest ways possible. The Chumash understanding and culture-based respect for Nature comes from their long and profound relationships with coastal marine ecosystems."[21]

The proposed CHNMS reflects another potential example of how conservation policy can make strides toward increasing inclusivity for ecosystems, people, and wildlife by incorporating important perspectives like TEK and compassionate conservation into its frameworks. While the process to designate the CHNMS is far from complete, it continues to make steady progress and will hopefully continue to do so.

Best Practices for Wildlife Corridors: A New Ethical Standard for Incorporating Compassion, Empathy, and Traditional Ecological Knowledge

Overall, several patterns have emerged over the course of the book: First, there is no single, perfect solution for implementing wildlife corridors in the Anthropocene, but that does not mean we ought not to try and move the needle in a productive direction. Second, it is possible to outline some best practices for corridor design that are grounded not only in current research and recommended practices in corridor design but also in emerging theories related to compassionate conservation, empathy, and TEK. Next, we can look at a range of case examples and see that it is possible to incorporate many of these practices within the scope of a corridor project; however, because corridor projects vary so much contextually, no single project will incorporate all of these practices or will even find all of the practices fully relevant. Moreover, there may be times when corridor projects must make concessions that involve biodiversity offsetting or prioritizing the needs of a flagship or keystone species first and foremost. In such cases, critical awareness and communication with decision-makers become that much more important to make sure that all perspectives are being explored and considered. Taken together, such best practices take shape as follows:

Best Practices for Wildlife Corridors: Integrating Conservation, Compassion, and Traditional Ecological Knowledge with Corridor Ecology

Best Practices Related to Current Research in Corridor Ecology

- As Anderson and Jenkins recommend, corridor design should focus on connecting patches of habitat that were previously connected prior to creating new disturbances in the landscape.[22]
- Gauging the appropriate width and habitat quality requires knowledge and observation of the species that must traverse the corridor in question.[23]
- Prior to creating new corridors, the existing ones should be evaluated and utilized when possible. This helps minimize habitat disturbances and reduce stress on species and landscapes by reducing any unnecessary human interventions.[24]
- Corridor design should identify and preserve currently existing natural corridors in order to help preserve water quality and maintain biodiversity.[25]
- Avoid expanses of landscape longer than one mile without also integrating nodes or endpoints like habitat patches or other structures, and incorporate redundant connections through the use of alternate pathways.[26]
- Corridor conservation should aim to identify strategies for preventing climate change impacts. These may include identifying specific habitats, communities, and populations that will be affected by climate change; mapping and tracking areas that will be most impacted by extreme weather or natural disasters; and distinguishing human-made impacts from natural changes in the physical environment.
- When designing movement corridors in response to climate change, Anderson and Jenkins also note that corridors should create a wide network that affords migration in multiple directions, in order to provide greater amounts of habitat and larger movement pathways for species with "relatively low mobility, such as many plants and terrestrial invertebrates."[27]

Overview: Best Practices Related to Empathy, Compassionate Conservation, and TEK

- Wildlife corridor projects should be grounded not only in current research in corridor ecology and recommended practices in corridor design but also in emerging theories related to compassionate conservation, empathy, and TEK.

- Wildlife corridors should be conceptualized from a mindset of compassionate conservation and should account for TEK whenever it is relevant and feasible to do so.
- In doing so, planners should solicit input from multiple stakeholders, such as biologists, state and tribal agencies, nonprofit organizations, local grassroots and public groups, and private landholders, whenever relevant and possible.

Best Practices Related to Empathy and Compassionate Conservation

- Wildlife corridor projects should account for the lives of all species that may utilize or migrate through specific habitats and seek to minimize any harm to wildlife to the extent possible, regardless of the purpose behind the action.
- Knowledge of animal behavior, in addition to expert opinion alone, is integral to understanding the environmental impacts on functional connectivity, or how species will make use of wildlife corridors.
- In order to best engage in compassionate conservation, corridor design should account not only for flagship and keystone species but also, ideally, for the needs of as many species in the habitat as possible, for what serves as a corridor for one species may not necessarily work as well for another.
- Knowledge of the needs of all wildlife affected by a corridor project can also better account for a compassionate conservation approach that avoids bias.
- To plan and design a wildlife corridor thus requires knowledge of all impacted species apart from solely umbrella species; this depth of knowledge can then arguably help foster greater empathy for a wider range of species.
- If and when corridor projects must make concessions that involve biodiversity offsetting or prioritizing the needs of a flagship or keystone species, first and foremost, critical awareness of the needs of all affected beings, as well as impacts to land, both ecological and cultural, should be considered from multiple perspectives, including those informed by TEK, whenever relevant.
- To incorporate the principles of compassionate conservation and empathy in corridor design also means understanding the behaviors, patterns, and requirements of the specific species that would use the corridor. The more that is known about a species, including knowledge of their social organization and dispersal patterns, the greater the likelihood that they will locate, make use of, and successfully pass through a corridor. Such

ideas also reflect the ways that compassionate conservation can align with the principles of corridor ecology.
- Remember that wildlife live with and among us and often inhabit urban ecologies. They are, in essence, stakeholders in corridor projects. In order to work toward coexistence with people and wildlife, we must seek to understand and advocate for their perspective as well.
- It is also possible to help foster connectivity at the individual level, by making relatively small adjustments to our ways of living. Such adjustments might be as simple as leaving small gaps in fencing, such that wildlife does not get trapped in a specific location and can easily find their way through neighborhoods and public spaces, or turning off lights during certain hours and times of the year when migratory birds are traveling through certain areas. Individual actions can work collectively to achieve connectivity and foster coexistence.

Best Practices Related to Traditional Ecological Knowledge (TEK)

- TEK is a valuable source of cultural and ecological knowledge and ought to be incorporated into connectivity projects when it is relevant and feasible to do so.
- Science and TEK are complementary to one another; where conservation science can identify the places that wildlife requires to thrive and migrate, protecting such places requires forming relationships with and learning from local peoples.
- The goals and approaches of connectivity projects should account for the needs and best interests of both people and wildlife whenever possible, and so corridor planning must include research such as stakeholder analyses, which can help paint a holistic picture of the various social, cultural, and economic contexts related to the habitats in question.
- Inclusivity toward and communication with all stakeholders and local peoples and communities has emerged as a critical component of land management and decision-making. Ongoing dialogue among all stakeholders is critical to the success and well-being of reconnected landscapes.

The practices outlined above may serve as guidelines for current and future corridor research and connectivity projects. And again, while it may be unrealistic to assume that the perfectly carried out connectivity project exists, and while we continue to learn from ongoing corridor research and projects, I suggest that an approach informed by compassion and TEK, and an interest in coexistence among all beings, nonhuman and human, can help forge a productive path forward for corridor ecology. It is my hope that this book has done just that, and that it can help extend the conversation and generate new ideas about inclusive practices in corridor design and planning. Wildlife corridors are, in essence, about coexistence, and as we move further into the age of the Anthropocene—an era characterized by unabating climate crisis and human development, it will become increasingly more necessary to learn to coexist with and among our nonhuman kin.

Notes

1. Conservation Corridor, "Corridor FAQ."
2. Anderson and Jenkins, *Applying Nature's Design*, 43.
3. Bekoff, *Rewilding*, 9.
4. Bekoff, *Rewilding*, 9.
5. Chumash National Marine Sanctuary, "About."
6. CHNMS, "About."
7. CHNMS, "About."
8. Foster-Frau, "Tribe Fights to Preserve."
9. Foster-Frau, "Tribe Fights to Preserve."
10. Foster-Frau, "Tribe Fights to Preserve."
11. Foster-Frau, "Tribe Fights to Preserve."
12. CHNMS, "About."
13. CHNMS, "About."
14. Hilty et al., *Corridor Ecology*, 235.
15. Hilty et al., *Corridor Ecology*, 235.
16. National Marine Sanctuaries, "Proposed Chumash Heritage National Marine Sanctuary."
17. National Marine Sanctuaries, "Proposed Chumash Heritage National Marine Sanctuary."
18. CHNMS, "An Area in Need of Protection."
19. National Marine Sanctuaries, "Proposed Chumash Heritage National Marine Sanctuary."
20. CHNMS, "An Area in Need of Protection."
21. CHNMS, "An Area in Need of Protection."
22. Anderson and Jenkins, *Applying Nature's Design*, 43.
23. Anderson and Jenkins, *Applying Nature's Design*, 43.
24. Anderson and Jenkins, *Applying Nature's Design*, 43.
25. Anderson and Jenkins, *Applying Nature's Design*, 43.
26. Anderson and Jenkins, *Applying Nature's Design*, 43.
27. Anderson and Jenkins, *Applying Nature's Design*, 35.

BIBLIOGRAPHY

Ament, Robert, Clevenger, Anthony, and van der Ree, Rodney, Eds. *Addressing Ecological Connectivity in the Development of Roads, Railways and Canals.* IUCN WCPA Technical Report Series No. 5. Gland, Switzerland: IUCN, 2023, Web. 16 Sept. 2023. https://portals.iucn.org/library/sites/library/files/documents/PATRS-005-En.pdf.

Anderson, Anthony B. and Jenkins, Clinton N. *Applying Nature's Design: Corridors as a Strategy for Biodiversity Conservation.* New York: Columbia UP, 2006.

Anderson, Jennifer. "Audubon to File Suit over Bird 'Slaughter.'" *Portland Tribune*, 31 Mar. 2015. Web. 18 June 2016.

Audubon. "Lights Out: Providing Safe Passage for Nocturnal Migrants." *National Audubon Society*, 2023. Web. 2 Sept. 2023. https://www.audubon.org/lights-out-program.

Austin, John M., Slesar, Chris, and Hammond, Forrest M. "Strategic Wildlife Conservation and Transportation Planning: The Vermont Experience." *Safe Passages: Highways, Wildlife, and Habitat Connectivity.* Eds. Beckmann, Jon P., Clevenger, Anthony P., Huijser, Marcel P., and Hilty, Jodi A. Washington: Island Press, 2010: 239–256.

Banner Baird, Joel. "New Monkton Salamander Crossing Saves Hundreds." *Burlington Free Press*, 28 Mar. 2016. Web. 13 Aug. 2023. https://www.burlingtonfreepress.com/story/news/2016/03/28/hundreds-saved-new-vermont-salamander-crossing/82336084/.

Barnett, Cynthia. "Hawaii Is Now Home to an Ocean Reserve Twice the Size of Texas." *National Geographic*, 26 Aug. 2016. Web. 21 May 2020. http://news.nationalgeographic.com/2016/08/obama-creates-world-s-largest-park-off-hawaii/.

Beckmann, Jon P. and Hilty, Jodi A. "Connecting Wildlife Populations in Fractured Landscapes." *Safe Passages: Highways, Wildlife, and Habitat Connectivity.* Eds. Beckmann, Jon P., Clevenger, Anthony P., Huijser, Marcel P., and Hilty, Jodi A. Washington: Island Press, 2010: 3–16.

Bekoff, Marc. *The Animal Manifesto: Six Reasons for Expanding Our Compassion Footprint.* Novato: New World Library, 2010.

Bekoff, Marc. *Rewilding Our Hearts: Building Pathways of Compassion and Coexistence.* Novato: New World Library, 2014.

Berkes, Fikret. *Sacred Ecology.* Third Edition. New York: Routledge, 2012.

Borda-de-Água, Luís, Barrientos, Rafael, Beja, Pedro, and Pereira, Henrique M., Eds. *Railway Ecology.* New York: Springer, 2017.

BSG Ecology. "East West Rail Ecology Mitigation." *BSG Ecology*, 2023. Web. 24 Aug. 2023. https://bsg-ecology.com/portfolio_page/east-west-rail-ecology-mitigation/.

Chaney, Rob. "Research on Highway 93 Wildlife Crossings Nearly Complete." *Missoulian*, 20 June 2013. Web. 21 May 2020. https://missoulian.com/lifestyles/recreation/

research-on-highway-93-wildlife-crossings-nearly-complete/article_aefa9130-d963-11e2-a600-001a4bcf887a.html.

Chiltern Railways. "Game-changing New Railway Wins Leading Industry Award." *Chiltern Railways*, 18 July 2017. Web. 21 May 2020. https://www.chilternrailways.co.uk/news/game-changing-new-railway-wins-leading-industry-award.

Christy, Alexandra and DiGirolamo, Mike. "Wildlife Crossings Built with Tribal Knowledge Drastically Reduce Collisions." *Video. Mongabay: News & Inspiration from Nature's Frontline*, 18 Nov. 2022. Web. 17 Jan. 2023. https://news.mongabay.com/2022/11/video-wildlife-crossings-built-with-tribal-knowledge-drastically-reduce-collisions/.

Chumash National Marine Sanctuary. "About the Proposed Chumash Heritage Sanctuary." *Chumash National Marine Sanctuary*, 2023. Web. 11 Sept. 2023. https://chumashsanctuary.org/about/.

Chumash National Marine Sanctuary. "An Area in Need of Protection." *Chumash National Marine Sanctuary*, 2023. Web. 11 Sept. 2023. https://chumashsanctuary.org/purpose/.

Cohen, Jeffrey Jerome. *Stone: An Ecology of the Inhuman*. Minneapolis: U of Minnesota P, 2015.

Connecticut Department of Energy and Environmental Protection (CT DEEP). "Long Island Sound Blue Plan." *Hartford: CT.gov*, 30 Dec. 2022. Web. 2 July 2023. https://portal.ct.gov/DEEP/Coastal-Resources/LIS-Blue-Plan/Long-Island-Sound-Blue-Plan-Home.

Conservation Corridor: Connecting Science to Conservation. "Corridor FAQ." *Michigan State University: Conservation Corridor*, 2023. Web. 5 July 2023. https://conservationcorridor.org/the-science-of-corridors/.

Conservation Planning. "Monkton Wildlife Crossing: Planning for Safe Passage During Salamander Migration. 2005–2015." *Conservation Planning*, 9 Aug. 2023. https://storymaps.arcgis.com/stories/84e9c986d22e4864b4c3b78660ca442e.

Corbett, Julia B. *Communicating Nature: How We Create and Understand Environmental Messages*. Washington, DC: Island Press, 2006.

The Cornell Lab of Ornithology. "Study Names Top Cities Emitting Light that Endangers Migratory Birds." *The Cornell Lab of Ornithology*, 1 Apr. 2019. Web. 29 Aug. 2023. https://mailchi.mp/cornell/release-study-lists-top-cities-where-lights-endanger-migratory-birds-1314385.

County of Santa Cruz Planning Department. "What Is a Riparian Corridor?" *County of Santa Cruz Planning Department*, 2023. Web. 27 July 2023. https://sccoplanning.com/PlanningHome/Environmental/Riparian/WhatisaRiparianCorridor.aspx.

Dapcevitch, Madison. "American Skyscrapers Kill an Estimated 600 Million Migratory Birds Each Year." *EcoWatch: Environmental News for a Healthier Planet and Life*, 10 Apr. 2019. Web. 21 May 2020. https://www.ecowatch.com/birds-killed-skyscrapers-light-pollution-2634222993.html.

eBird. "Oyster River Mouth, New Haven County, Connecticut, US." *eBird*, 2023. Web. 2 July 2023. https://ebird.org/hotspot/L158203.

Executive Office of the President: Office of Science and Technology Policy (OSTP) and Council on Environmental Quality (CEQ). "Memorandum: Indigenous Traditional Ecological Knowledge and Federal Decision Making." *Executive Office of the President*, 15 Nov. 2021. Web. 9 Sept. 2023. https://www.whitehouse.gov/wp-content/uploads/2021/11/111521-OSTP-CEQ-ITEK-Memo.pdf.

Falchi, Fabio, Cinzano, Pierantonio, Duriscoe, Dan, Kyba, Christopher C. M., Elvidge, Christopher D., Baugh, Kimberly, Portnov, Boris A., Rybnikova, Nataliya A., and Furgoni, Riccardo. "The New World Atlas of Artificial Night Sky Brightness." *ScienceAdvances*. 2, 6 (2016): 1–25.

Foster-Frau, Silvia. "Tribe Fights to Preserve California Coastline — And Its Own Culture." *The Washington Post*, 29 July 2023. Web. 17 Sept. 2023. https://www.washingtonpost.com/nation/2023/07/29/chumash-tribe-california-marine-sanctuary/.

Francis, Kirk Sr. "Honoring Traditional Ecological Knowledge Is Critical." *United South and Eastern Tribes, Inc*, 13 Dec. 2021. Web. 9 Sept. 2023. https://www.usetinc.org/general/honoring-traditional-ecological-knowledge-is-critical/.

Fritts, Rachel. "Avian Superhighways: The Four Flyways of North America." *American Bird Conservancy*, 16 May 2022. Web. 31 Aug. 2023. https://abcbirds.org/blog/north-american-bird-flyways/.

Golden Gate Bird Alliance. *Lights Out for Birds*. Berkeley: Golden Gate Bird Alliance, 2023. Web. 2 Sept. 2023. https://goldengateaudubon.org/conservation/make-the-city-safe-for-wildlife/learn-about-lights-out-san-francisco/.

Griffin, Andrea S., Callen, Alex, Klop-Toker, Kaya, Scanlon, Robert J., and Hayward, Matt W. "Compassionate Conservation Clashes with Conservation Biology: Should Empathy, Compassion, and Deontological Moral Principles Drive Conservation Practice?" *Frontiers in Psychology*. 11 (2020): 1–9.

Gruen, Lori. *Entangled Empathy: An Alternative Ethic for Our Relationships with Animals*. New York: Lantern Books, 2015.

Haq, Shiekh Marifatul, Pieroni, Andrea, Bussmann, Rainer W., Abd-ElGawad, Ahmed M., and El-Ansary, Hosam O. "Integrating Traditional Ecological Knowledge into Habitat Restoration: Implications for Meeting Forest Restoration Challenges." *Journal of Ethnobiology and Ethnomedicine*. 19, 33 (2023): 1–19.

Hauser, Christine. "Turn the Lights Out: Here Come the Birds." *The New York Times*, 10 Apr. 2021. Web. 17 Sept. 2023. https://www.nytimes.com/2021/04/10/us/bird-migration-lights-out.html?fbclid=IwAR2R2U-fLIjsojukK1EcoaDP5LKKVlfp-Yy9ppZK6dZiXl1iRWaYwt2FsK4.

Hilty, Jodi A., Keeley, Annika T.H., Lidicker, William Z. Jr., and Merenlender, Adina M. *Corridor Ecology*. Second Edition. Washington, DC: Island Press, 2019.

Hilty, Jodi A., Worboys, Graeme L., Keeley, Annika, Woodley, Stephen, Lausche, Barbara, Locke, Harvey, Carr, Mark, Pulsford, Ian, Pittock, James J., White, Wilson, Theobald, David M., Levine, Jessica, Reuling, Melly, Watson, James E.M., Ament, Rob, and Tabor, Gary M., Eds. *Guidelines for Conserving Connectivity through Ecological Networks and Corridors*. Best Practice Protected Area Guidelines Series No. 30. Gland, Switzerland: IUCN, 2020. Web. 16 Sept. 2023. https://portals.iucn.org/library/sites/library/files/documents/PAG-030-En.pdf.

Hilty, Jodi A. and Zenkewich, Kelly. "The Future of Large Landscape Conservation Begins with Indigenous Communities." *High Country News*, 1 Nov. 2022. Web. 13 July 2023. https://www.hcn.org/issues/54.11/public-lands-the-future-of-large-landscape-conservation-begins-with-indigenous-communities.

Hirschfeld Davis, Julie. "Obama to Create World's Largest Marine Reserve Off Hawaii." *The New York Times*, 26 Aug. 2016. Web. 21 May 2020. https://www.nytimes.com/2016/08/26/us/politics/obamas-action-will-create-largest-marine-reserve-on

-earth.html?action=click&contentCollection=Politics&module=RelatedCoverage&region=EndOfArticle&pgtype=article&_r=0.

Horton, Kyle G., Nilsson, Cecilia, Van Doren, Benjamin M., La Sorte, Frank A., Dokter, Adriaan M., and Farnsworth, Andrew. "Bright Lights in the Big Cities: Migratory Birds' Exposure to Artificial Light." *Frontiers in Ecology and the Environment.* 17, 4 (2019): 209–214.

Huijser, Marcel P., Camel-Means, Whisper, Fairbank, Elizabeth R., Purdum, Jeremiah P., Allen, Tiffany D.H., Hardy, Amanda R., Graham, Jonathan, Begley, James S., Basting, Pat, and Becker, Dale. "US 93 North Post-Construction Wildlife-Vehicle Collision and Wildlife Crossing Monitoring on the Flathead Indian Reservation Between Evaro and Polson, Montana, FHWA/MT-16-009/8208: Final Report Prepared for the Montana Department of Transportation by the Western Transportation Institute." Nov. 2016. Web. 31 July 2023. https://www.mdt.mt.gov/other/webdata/external/research/docs/research_proj/wildlife_crossing/phaseii/PHASE_II_FINAL_REPORT.pdf.

Huijser, Marcel P. and McGowen, Pat T. "Reducing Wildlife-Vehicle Collisions." *Safe Passages: Highways, Wildlife, and Habitat Connectivity.* Eds. Beckmann, Jon P., Clevenger, Anthony P., Huijser, Marcel P., and Hilty, Jodi A. Washington: Island Press, 2010: 51–74.

International Union for Conservation of Nature (IUCN). "Biodiversity Offsets." *IUCN: Issues Brief,* Sept. 2016. Web. 4 Sept. 2023. https://www.iucn.org/resources/issues-brief/biodiversity-offsets.

Kershner, Jessi. "Incorporating Climate Change Adaptation into the Papahānaumokuākea Marine National Monument Management Plan." *Climate Adaptation Knowledge Exchange (CAKE),* 15 Nov. 2021. Web. 16 Aug. 2023. https://www.cakex.org/case-studies/incorporating-climate-change-adaptation-papah%C4%81naumoku%C4%81kea-marine-national-monument-management-plan.

Kimmerer, Robin Wall. *Braiding Sweetgrass.* Minneapolis: Milkweed Editions, 2013.

Kolb Noyes, Amy. "First Salamander Sighted in Monkton's Amphibian Underpass." *Vermont Public,* 28 Mar. 2016. Web. 13 Aug. 2023. https://www.vermontpublic.org/vpr-news/2016-03-28/first-salamander-sighted-in-monktons-amphibian-underpass.

Kousky, Carolyn. "Chicago Skyscrapers Go Dark for Migratory Birds: A Building Less Bright." *Terrain.org: A Journal of the Built & Natural Environments* 15 (2004). https://www.terrain.org/articles/15/kousky.htm.

Lamb, Clayton T., Willson, Roland, Richter, Carmen, Owens-Beek, Naomi, Napoleon, Julian, Muir, Bruce, McNay, R. Scott, Lavis, Estelle, Hebblewhite, Mark, Giguere, Line, Dokkie, Tamara, Boutin, Stan, and Ford, Adam T. "Indigenous-led Conservation: Pathways to Recovery for the Nearly Extirpated Klinse-Za Mountain Caribou." *Ecological Applications: Ecological Society of America,* 23 Mar. 2022. Web. 16 July 2023. https://doi.org/10.1002/eap.2581.

Lewis Creek Association. "The Monkton Road Wildlife Crossing Project: Ensuring the Survival of a Critical Amphibian Migration Corridor." Brochure. *Lewis Creek Association,* Web. 9 Aug. 2023. https://static1.squarespace.com/static/57d1b980d482e9f1f1079811b/t/58ab4e0fff7c50f538f4de4b/1487621649344/Monkton_Wildlife_and_Amphibian_Crossing_Project__2013.pdf.pdf.

Lewis Creek Association. "Monkton Wildlife Crossing." *Lewis Creek Association,* Web. 21 May 2020. http://www.lewiscreek.org/monkton-wildlife-crossing.

Li-Chee-Ming, Samantha. "Preserving Montana's Biodiversity." *WildMontana.org*, 2015. Web. 7 Jan. 2022. https://storymaps.arcgis.com/stories/fa2966954ec446f4b754025 5812117f0.

Lorimer, Jamie. *Wildlife in the Anthropocene: Conservation after Nature*. Minneapolis: U of MN P, 2015.

MacKay, Paula. "Where Compassionate Conservation Meets Rewilding, with Marc Bekoff." *Wildlands Network*, 30 Apr. 2018. Web. 22 May 2020. https://wildlandsnetwork .org/blog/compassionate-conservation-marc-bekoff/.

McGregor, Deborah. "Coming Full Circle: Indigenous Knowledge, Environment, and Our Future." *American Indian Quarterly*. 28, 3&4 (2004): 385–410.

McNaughton, Angelique. "UDOT Completes Utah's Largest Wildlife Crossing at Parleys Summit." *Park Record*, 14 Dec. 2018. https://www.parkrecord.com/news/udot -completes-utahs-largest-wildlife-crossing-at-parleys-summit/.

Meyer, Ninon F.V., Moreno, Ricardo, Reyna-Hurtado, Rafael, Signer, Johannes, and Balkenhol, Niko. "Towards the Restoration of the Mesoamerican Biological Corridor for Large Mammals in Panama: Comparing Multi-Species Occupancy to Movement Models." *Movement Ecology*. 8, 3 (2020): 1–14.

Monterey Bay Aquarium. "Sea Otter." *Monterey Bay Aquarium*, 2023. Web. 30 July 2023. https://www.montereybayaquarium.org/animals/animals-a-to-z/sea-otter.

National Ocean Service: Office of National Marine Sanctuaries, National Oceanic and Atmospheric Administration. "Papahānaumokuākea Marine National Monument." *National Ocean Service*, 20 Feb. 2019. Web. 21 May 2020. https://www. papahanaumokuakea.gov/wheritage/refuge.html.

National Ocean Service: Office of National Marine Sanctuaries, National Oceanic and Atmospheric Administration. "Papahānaumokuākea: A Sacred Name, A Sacred Place." *National Ocean Service*, 20 Apr. 2023. Web. 16 Aug. 2023. https://www .papahanaumokuakea.gov/new-about/name/.

National Marine Sanctuaries. "Proposed Chumash Heritage National Marine Sanctuary." *National Ocean Service*, 2023. Web. 13 Sept. 2023. https://sanctuaries.noaa .gov/chumash-heritage/.

National Park Foundation. "Saint Croix National Scenic Riverway." *National Park Foundation*, 2023. Web. 11 July 2023. https://www.nationalparks.org/explore/parks /saint-croix-national-scenic-riverway.

The Nature Conservancy. "Filming the Ghost Cat." *The Nature Conservancy*, 2023. Web. 12 July 2023. https://preserve.nature.org/page/87254/data/1?&supporter.appealCode =AHOMA210801G0XXX01&src=e.dfr.eg.x.pnthr2.n.n.sas&en_txn8=NewSch. REEMSA2108NPNZFEBE15Z01-ZZZZZ-DGRT&lu=87daa177-12ec-4496-b0ec -d5f6381f2bb6&ea.campaigner.email=lZQgIcBlj0rCC0SyLpfpJsqntdjf%2FMKW1 va7QlgcnA0%3D.

The Nature Conservancy. "Florida Panthers: Crossing the Caloosahatchee." *The Nature Conservancy*, 2 Sept. 2018. Web. 12 July 2023. https://www.nature.org/en-us/about -us/where-we-work/united-states/florida/stories-in-florida/florida-panther-kittens -north-of-caloosahatchee/.

Network Rail. "Oxford to London Marylebone Rail Link Guards Wildlife Habitats and Protected Species." *Network Rail*, 12 May 2016. Web. 21 May 2020. https://www.net workrailmediacentre.co.uk/news/oxford-to-london-rail-link-guards-wildlife-habitats -and-protected-species.

Orange Historical Society. "History." *Orange, CT*, 2023. Web. 2 July 2023. https://orangehistory.org/town_of_orange_history.htm.

Papahānaumokuākea Marine National Monument. "Integrating Native Hawaiian Culture into Management of Papahānaumokuākea." *Papahānaumokuākea Marine National Monument*, 21 June 2021. Web. 9 Sept. 2023. https://www.papahanaumokuakea.gov/new-news/2021/06/21/maikapomai/.

Pacific Birds: Habitat Joint Venture. "Birds of Papahānaumokuākea." *Pacific Birds*, 5 Sept. 2016. Web. 16 Aug. 2023. https://pacificbirds.org/2016/09/birds-of-papahanaumokuakea/.

People's Way Partnership. "US 93 North Wildlife Passages - Montana." People's Way Wildlife Crossing Highway 93 North Montana: Montana State University Western Transportation Institute, 2019. Web. 21 May 2020. https://westerntransportationinstitute.org/programs/road-ecology/peoples-way-partnership/.

Pierce, Scott D. "New $5 Million, Animals-Only Overpass at Parleys Summit Is Saving Wildlife (and Drivers) Already." *The Salt Lake Tribune*, 20 June 2019. https://www.sltrib.com/news/2019/06/20/new-million-animals-only/.

Planet Forward Staff. "2017 Summit: Stories That Last: Native American Traditions of Storytelling with Dr. Robin Kimmerer." *Planet Forward*, 18 Apr. 2017. Web. 14 Sept. 2023. https://www.planetforward.org/idea/2017-summit-stories-that-last-native-american-traditions-of-storytelling-with-dr-robin-kimmerer.

Plumwood, Val. *Feminism and the Mastery of Nature*. New York: Routledge, 1993.

Popp, Jesse N. and Boyle, S.P. "Railway Ecology: Underrepresented in Science?" *Basic and Applied Ecology*. 19 (2017): 84–93.

Popp, Jesse N. and Hamr, Josef. "Seasonal Use of Railways by Wildlife." *Diversity*. 10 (2018): 1–10.

Pratt, Beth. "How a Lonely Cougar in Los Angeles Inspired the World." *TEDx Talks*, 22 Feb. 2016. Web. 13 May 2020. https://www.youtube.com/watch?v=pMO8-f70nFY.

Pressley Associates, Inc. *The Emerald Necklace Parks: Master Plan*. Commonwealth of Massachusetts, Executive Office of Environmental Affairs, 2001. https://ia902601.us.archive.org/3/items/emeraldnecklacep00walm/emeraldnecklacep00walm.pdf.

Railway Electrical Services (RES). "Wolvercote Tunnel." *Railway Electrical Services Limited*, 2023. Web. 24 Aug. 2023. https://www.railwayelectricalservices.co.uk/wolvercote-tunnel/.

Railway Technology. "Railways and Wildlife: Coexisting in an Age of Expansion." *Railway Technology*, 16 June 2016. Web. 21 May 2020. https://www.railway-technology.com/features/featurerailways-and-wildlife-coexisting-in-an-age-of-expansion-4923598/.

Ramos, Seafha C. (Yurok/Karuk), Shenk, Tanya M., and Leong, Kirsten M. "Introduction to Traditional Ecological Knowledge in Wildlife Conservation: Natural Resource Report NPS/NRSS/BRD/NRR—2016/1291." *U.S. Department of the Interior, National Park Service: Natural Resource Stewardship and Science*. Fort Collins, CO, Aug. 2016. Web. 7 Sept. 2023. https://wildlife.org/wp-content/uploads/2022/05/Ramos-et-al.2016.Introduction-to-Traditional-Ecological-Knowledge-in-Wildlife-Conservation.pdf.

Santos, Sara M., Carvalho, Filipe, and Mira, António. "Current Knowledge on Wildlife Mortality in Railways." *Railway Ecology*. Eds. Borda-de-Água, Luís, Barrientos, Rafael, Beja, Pedro, and Pereira, Henrique M. New York: Springer, 2017: 11–22.

Seeger, Eric. "The Panther's Path." *The Nature Conservancy*, 31 Oct. 2019. Web. 13 July 2023. https://www.nature.org/en-us/magazine/magazine-articles/florida-panther-corridor/.

Shilling, Dan. "Introduction: The Soul of Sustainability." *Traditional Ecological Knowledge: Learning from Indigenous Practices for Environmental Sustainability*. Eds. Nelson, Melissa K. and Shilling, Dan. Cambridge: Cambridge UP, 2021: 3–14.

Shinn, Lora. "Montana's Wildlife Needs Safer Crosswalks." *National Resources Defense Council*, 13 Mar. 2019. Web. 7 Jan. 2022. https://www.nrdc.org/stories/montanas-wildlife-needs-safer-crosswalks.

Slesar, Chris. "Movin' Lizards: 'Hey. You Movin' Lizards? Are the Lizards Out Tonight?'" *The Orianne Society: Science. Conservation. Education*. Winter (2020): 16–23. Web. 9 Aug. 2023. https://static1.squarespace.com/static/57d1b980d482e9f1f107981b/t/5e7d258b7fa6e328446dd9d9/1585259982398/Moving-Lizards-Chris-Slesar.pdf.

Smith, Daniel S. and Hellmund, Paul Cawood. *Ecology of Greenways: Design and Function of Linear Conservation Areas*. Minneapolis: U of Minnesota P, 1993.

State of Hawai'i: Division of Aquatic Resources. "Papahānaumokuākea Marine National Monument." *State of Hawai'i*, 2023. Web. 14 Sept. 2023. https://dlnr.hawaii.gov/dar/marine-managed-areas/papahanaumokuakea-marine-national-monument/.

Stowe Land Trust. "Putting the Shutesville Hill Wildlife Corridor on the Map." *Stowe Land Trust*, 22 Dec. 2017. Web. 5 Apr. 2019. https://www.stowelandtrust.org/news/post/news-putting-the-shutesville-hill-wildlife-corridor-on-the-map.

Terrascope. "Boston Background." *MIT*, 2023. Web. 12 July 2023. https://terrascope2024.mit.edu/?page_id=752

Tobias, Michael Charles. "Compassionate Conservation: A Discussion from the Frontlines with Dr. Marc Bekoff." *Forbes*, 9 May 2013. Web. 8 Mar. 2017. https://www.forbes.com/sites/michaeltobias/2013/05/09/compassionate-conservation-a-discussion-from-the-frontlines-with-dr-marc-bekoff/3/#2c75be34444a.

Two Countries, One Forest. Web, 2019. 11 July 2023. https://2c1forest.org/.

U.S. Department of the Interior: Office of Congressional and Legislative Affairs. "Antiquities Act." *U.S. Department of the Interior: Office of Congressional and Legislative Affairs*, 2023. Web. 16 Aug. 2023. https://www.doi.gov/ocl/antiquities-act.

U.S. Fish & Wildlife Service. "Traditional Ecological Knowledge Fact Sheet (Feb. 2011)." *U.S. Fish & Wildlife Service*, Feb. 2011. Web. 20 May 2020. https://www.fws.gov/nativeamerican/pdf/tek-fact-sheet.pdf.

UTS. "What Is Compassionate Conservation?" *Centre for Compassionate Conservation, University of Technology Sydney*, 2015. Web. 14 Sept. 2023. https://www.uts.edu.au/research-and-teaching/our-research/centre-compassionate-conservation/about-us/what-compassionate.

@UtahDWR. "We're Excited to see #wildlife Using the New Parleys Summit Overpass!" *Twitter*, 20 June 2019. https://twitter.com/UtahDWR/status/1141716080206868480.

Vermont Fish & Wildlife Department. "Blue-Spotted Salamander." *Agency of Natural Resources: Vermont Fish & Wildlife Department, State of Vermont*, 2023. Web. 13 Aug. 2023. https://vtfishandwildlife.com/learn-more/vermont-critters/amphibians/blue-spotted-salamander.

Vermont Natural Resources Council (VNRC). "Wildlife Corridor Protection." *VNRC*, Web. 5 Apr. 2019. https://vnrc.org/community-planning-toolbox/issues/wildlife-corridor-protection/.

Watson, Julia. *Lo—TEK: Design by Radical Indigenism*. Los Angeles: TASCHEN, 2020.

Yellowstone to Yukon Conservation Initiative. "About." *Yellowstone to Yukon Conservation Initiative*, 2023. Web. 13 July 2023. https://y2y.net/about/.

Yellowstone to Yukon Conservation Initiative. "History." *Yellowstone to Yukon Conservation Initiative*, 2023. Web. 13 July 2023. https://y2y.net/about/vision-mission/history/.

Yellowstone to Yukon Conservation Initiative. "Our Work." *Yellowstone to Yukon Conservation Initiative*, 2023. Web. 13 July 2023. https://y2y.net/work/.

INDEX

Note: Bold page number refer to figures; Page numbers followed by "n" denote endnotes.

aerial corridors, in urban environments 83–88
aero corridor 10, 80–81, 83
Ament, Robert 7, 62
amphibian crossing tunnels: *see* Monkton Road Wildlife Crossing Project
Anderson, Anthony B. 8, 14, 17–18, 32–33, 37, 39, 106
"Animals' Bridge" 57, 58, **61**
Anthropocene 3, 8, 13, 109; wildlife corridors in 105
anthropocentric perspectives 7
anti-humanism 44
area-based conservation measures (OECMs) 14
artificial light at night (ALAN) 85
assisted migration 15
avian superhighways 84

badgers 79–80
"Bat Roosting Site" 81
Bekoff, Marc 5, 31, 44–45, 47, 49–50, 61, 65, 101
Berkes, Fikret 3, 6
best practices, for wildlife corridors: ethical standard for incorporating compassion, empathy, and TEK 105; related to empathy and compassionate conservation 107–8; related to traditional ecological knowledge 108–9; research in corridor ecology 106
Big Cypress National Preserve 23
biodiversity 8, 13, 49, 101–2; balancing needs 35–36; challenges of offsetting 35–36; conservation of 17, 65; ecosystems and 77; vibration and noise on 78

Bird Collision Monitor and Rescue Project 86–87
bird conservation 85
BirdCast program 86
blue-spotted salamander (*Ambystoma laterale*) **70**, 70–71
Borda-de-Água, Luís 77

California Native American Tribes 104
Caloosahatchee Ecoscape 33–34
Caloosahatchee River 23
caribou habitat 25–26
Centre for Compassionate Conservation 43
Channel Islands National Marine Sanctuary 104
Chasis, Sarah 96
Chiltern Railways: case of 35; compassionate conservation and railway ecology 81; habitat reconstruction project 78–79; reconstruction project 49, 83; upgrade project 102
Chumash Heritage National Marine Sanctuary (CHNMS) 103–5
climate change: accounting for 95–96; monument to 96; unpredictability of 97
climate crisis 22, 37; and human development 109
coastal marine ecosystems 105
Cohen, Jeffrey Jerome 7
communication 40; critical awareness and 105; with stakeholders and local peoples and communities 52
compassionate conservation 3–5, 31, 43–44, 97, 99, 101; best practices related to 107–8; critiques of 48–49; engage

in 36–37; ideas of 48; and Lights Out campaigns 86–88; principles of 31, 36, 48, 80, 107; and railway ecology 81; rewilding 45; at work 22–24
compassionate conservationists 48
complex relationships 45–46
Confederated Salish and Kootenai Tribes (CSKT) 51, 57
connectivity: aerial corridors relative to wildlife 102; challenges of 32–33; and scale 16–17; science-based project 49; structural and functional 15–16; understanding of 13–15, **14**
connectivity goals: account for needs of stakeholders 33–35; challenges of biodiversity offsetting 35–36; climate considerations 37; engage in compassionate conservation approach 36–37
connectivity projects 4, 17, 26; conceptualization, design, and management of 31; goals and approaches of 108
conservation, "Indigenous worldview" about 25
conservation biology 47, 50
Conservation Corridor 13
conservation management strategies 15
conservation projects: critiques of compassionate conservation 48–49; and different ideas about animal rights 47; middle ground in nuances of engagement and critical awareness 49–50
conservation science 34
Cook, Tanith 79, 81
Coral Reef Ecosystem Division, of NOAA Pacific Islands Science Center 95
Coral Reef Ecosystem Integrated Observing System 95
Coral Reef Watch 95
Corbett, Julia 30–31
Cornell Lab of Ornithology 83, 85
Cornell Ornithology Lab 86
corral-style fencing 18, **19**
corridors: conservation 106; ecology, best practices related to research in 106; habitat 14; *see also* wildlife corridors
Cross Industry Partnership Award 81
cultural transmission 3

Dering-Dibru Saikhowa Elephant Corridor 52
dispersal ecology 37

"do no harm" philosophy 48–49
drift fencing 72

eBird app 1
Ebisui, Edwin, Jr. 93
ecological connectivity 2–3, 13
ecological corridors 13–14; marine 15
ecological network for conservation 13–14
ecologistic attitudes 30
ecosystems: and biodiversity 77; management of 50; rationales for valuing 30–32
Emerald Necklace Park System 21
empathy 2; and compassionate conservation 31; distinguishing between sympathy and 5–6
entangled empathy 3, 5–6, 45–47, 99–101; best practices related to 107–8
Environmental Excellence Award 74
European Protected Species (EPS) 79
European Railway Agency 77–78
Everglades National Park 23
evidence-based approaches 70

Falchi, Fabio 84
Federal Highways Administration (FHWA) 57, 74
fencing 29; afford both structural and functional connectivity 62–63; effectiveness of 72
Field Museum, in Chicago 87
fishing culture 93, 96
flagship species 34, 46
Flathead Indian Reservation 45, 51, 57–58, 72, 102; wildlife crossing on 65; wildlife crossing structures 58, **59**; wildlife exclusion fences 58, **60**; wildlife jump-out/escape ramp 58, **61**; wildlife overpass 58, **61**
Florida Department of Transportation 23
Florida panther 22–24, 34
Florida Panther National Wildlife Refuge 23
fragmentation: of habitats 14, **14**; landscape 3
functional connectivity 15–16, 18, 20, 24, 64–65; wildlife fences afford both structural and 62–63

geographic information science (GIS) 16
Gillin, Stephanie 51
Greater Yellowstone Ecosystem 24
green infrastructure 22; and Florida panther 22–24

INDEX

green infrastructure corridor projects 22
greenways/open-space systems/greenbelts 21, **22**
Griffin, Andrea S. 48
grizzly bear 63–64
Gruen, Lori 5–6, 45–48

habitat reconstruction project, Chiltern Railways 78–79
habitats 13; conservation approaches 79–80; corridor design focus on connecting patches of 106; fragmentation of 14, **14**; linkages between 4; physical characteristics of 15; quality of 38
Haq, Shiekh Marifatul 52
Hawai'i Department of Land and Natural Resources 93
Hawaiian community 95, 97
Hawaiian Islands Bird Reservation 93
Hawaiian Islands National Wildlife Refuge 93
Highway 93, 57–58, 72; effectiveness of highway reconstruction 62; monitoring and usage 63–64; "road is a visitor, the" 64–66; stakeholders, governance, and the value of partnerships 64; wildlife crossing structures 58, **59**; wildlife exclusion fences 58, **60**; wildlife jump-out or escape ramp 58, **61**; wildlife overpass 58, **61**
highway reconstruction project 62–63
Hilty, Jodi A. 8, 18, 20, 25–26, 34, 83
Horton, Kyle 84–85, 87
Huijser, Marcel P. 66n5–8, 10, 12–17, 19–28
human development 2; Anthropocene era of 50; climate crisis and 109; forces of 30; impact on natural world 3
human-centric assumptions 7
human-designed connectivity projects 22
humanism 44
human-made underpass 33

Incashola, Tony 66
Indigenous peoples 51; subsistence and cultural values of 25; and TEK 66
Indigenous Traditional Ecological Knowledge (ITEK) 53
"Indigenous Traditional Ecological Knowledge and Federal Decision Making" 93, 104
Indigenous-led conservation agreements 25

Interagency Working Group on Indigenous Traditional Ecological Knowledge 53
ITEK: *see* Indigenous Traditional Ecological Knowledge

Jenkins, Clinton N. 8, 14, 17–18, 32–33, 37, 39, 106
John Hancock Building 86
jump-out/escape ramp 58, **61**, 64

Kauai island 91
keystone species 34, 46
Kimmerer, Robin Wall 7, 25, 36, 39, 52
Klinse-Za subpopulation of caribou 25–26
knowledge: of animal behavior 107; bodies of 53; of life-worlds 46; philosophies of 52
"Kumulipo, the" 94

LaDuke, Winona 51
landscape fragmentation 3
landscapes: climate crisis 37; connectivity on 18; functioning as corridors 20; wildlife and leverages 33
larger-scale corridor initiatives 16–17
Laysan Albatross 91, **92**
Laysan Finch 91
"learning curve" 62
Lewis Creek Association 71, 74
light pollution: conservation approaches aimed at reducing 85–86; defined as 84; impacts on migrating birds 84–85
lighting project 83
Lights Out campaigns 84, 86–88, 102
linear transport infrastructure (LTI) 3, 7
Long Island Sound 1
Lorimer, Jamie 7
LTI: *see* linear transport infrastructure

"Mai Ka Pō Mai" document 96–97
marine conservation project 103–5
marine protected areas (MPAs) 15, 92, 94, 96, 102; establishment of 104
marine systems, accounting for 102
MBC: *see* Mesoamerican Biological Corridor
McGregor, Deborah 51–52
Means, Whisper Camel 65
Memorandum of Agreement (MOA) 93
Mesoamerican Biological Corridor (MBC) 17–18

Meyer, Ninon F.V. 17
Michel, Paul 103
Midway Atoll 91
Midway Atoll National Wildlife Refuge 93
Midway Island 96
migrating birds: aerial corridors for 84; impacts of light pollution on 84–85; species of 83
Milky Way 84
Milne, Andy 79
Monkton Conservation Commission 71
Monkton Road Wildlife Crossing Project 74
Monkton Wildlife Crossing Tunnels 69–72, **70**, 102; crossing structures 72–73, **73**; tunnels as acts of compassionate conservation 73–74
Monkton-Vergennes Road 70
Montana Department of Transportation (MDT) 57
Montana State University 62
Monterey Bay National Marine Sanctuary 104–5
moralistic attitudes 30
MPAs: *see* marine protected areas

National Audubon Society 85, 87
National Marine Sanctuary Program 104–5
National Oceanic and Atmospheric Administration (NOAA) 53, 93, 103
Native American reservation 59–60
Native American tribes 66
Native communities 52–53
Native Hawaiian community 94, 96–97
natural disasters, unpredictability of 97
Natural England 36
Natural Resources Defense Council 96
naturalistic attitudes 30
nature 7; economy of 66
Nature Conservancy (TNC), The 23
New York City Audubon 85
nighttime-migrating birds 85
Nihoa Finch 91
Nihoa Millerbird 91
Niihau island 91
NOAA: *see* National Oceanic and Atmospheric Administration
Northern Chumash Tribal Council 103
Northern Rockies 64
Northwestern Hawaiian Islands (NWHI) 53, 91–92, 95
Northwestern Hawaiian Islands Marine National Monument 94–95
Northwestern Hawaiian Islands Native Hawaiian Cultural Working Group 95
"no-take" area 93

Obama, Barack 92, 96
Olmsted, Frederick Law 21
Orange, Connecticut 1–2
overpass 58, **61**, 63

Papahānaumokuākea Marine National Monument (PMNM) 15, 40, 53–54, 91, 102; accounting for climate change 95–96; boundary expansion of 92, 96; co-management structure of 94; creation of 92–93; Laysan Albatross colony on 91, **92**; management of 93–94; naming of 94–95
Parleys Summit wildlife crossing 29–31, 33, 100
Parren, Steve 71
partnerships, Highway 93 64
patches 14–15, 20–21, 33, 39, 101
Plumwood, Val 7
PMNM Management Plan 93–95, 97
principles of compassion 2

Railway Electrical Services (RES) 81
railway systems 77; accounting for needs of various species 80–81; approaches to habitat conservation 79–80; Chiltern Railways habitat reconstruction project 78–79; compassionate conservation and railway ecology 81; wildlife implications of roadways *vs.* 77–78
Ramos, Seafha C. 52
restoration, "Indigenous worldview" about 25
rewilding projects 44–45
riparian corridors 38
Road Ecology program 62
"road is a visitor, the" 64–66
roadside vegetation 20–21, **21**
roadways 3–4; wildlife implications of railway systems *vs.* 77–78
Roosevelt, Theodore 92–93

Saint Croix National Scenic Riverway 16
salamanders **70**, 70–71
Santos, Sara M. 78
Saulteau First Nations 25
Save People Save Wildlife 29

scale 2, 8, 10, 99; connectivity and 16–17
science-based project 49
Sears Tower 86
Selis Qlispe Culture Committee 66
Shilling, Dan 6
Slesar, Chris 69
spatial scale, of corridor 16–17
'Spirit of Place' approach 58
stakeholders 4, 7, 101; account for needs of 33–35; benefits to 31; challenges of land management 24; in corridor projects 108; as "a geography of hope" 24–26; governance, and value of partnerships 64; understanding obstacles and needs 32–33
State of Hawai'i Office of Hawaiian Affairs 93
State Road 80 22–24
structural connectivity 15–16, 18, 20; wildlife fences afford both functional and 62–63
sympathy 5–6

TEK: *see* traditional ecological knowledge
theory of entangled empathy 45
traditional ecological knowledge (TEK) 3, 4, 6–8, 39, 51–54, 58, 99–102; best practices related to 108–9; role in conservation management and planning 54
Trans-Canada highway 18, **20**
Tribal Nations 53
Two Countries, One Forest initiative 17

unintended corridors 20
United South and Eastern Tribes, Inc. (USET) 54
United States: scientific, technical, social, and economic advancements of 53
University of Technology Sydney 43
urban environments: aerial corridors in 83–88; impacts of light pollution on migrating birds 84–85
urban spaces 2
U.S. Endangered Species Act 91
U.S. Fish and Wildlife Service 8, 53, 93
U.S. Highway 93 North: reconstruction of 79
Utah, wildlife crossing in 29
Utah Department of Transportation (UDOT) 29–30
Utah Division of Wildlife Resources 30
utilitarian attitudes 30

Vermont Agency of Transportation (VTrans) 69, 74
Vermont Reptile and Amphibian Atlas Project 71, 74
Vital Ground 64

West Moberly First Nations 25
Western Pacific Regional Fishery Management Council 93
Western Science 8
Western Transportation Institute (WTI) 59, 62
White, Germaine 66
White House Council on Environmental Quality (CEQ) 53, 104
White House Office of Science and Technology Policy (OSTP) 53, 104
white-tailed deer 17, 63
wildlife: and leverages 33; overpass crossing 18, **20**; rationales for valuing 30–32; underpass crossing 18, **19**
wildlife corridors 2, 16, 30, 100, 106–7; in Anthropocene 105; balancing multiple perspectives and sources of knowledge 40; benefits for nonhuman species 17–18; challenges of compassionate coexistence between wildlife and people 7; communication with stakeholders and local peoples and communities 52; and connectivity projects 99; creation of 4; design and creation of 3; design elements and considerations 38–39; design guidelines for 39, 101; as green infrastructure 22; need for 101; needs of flagship/keystone species 34; practice of 43; public support for 46; types of 18, **19–22**, 20–21; value of partnerships for progress of 64; *see also* corridors
wildlife crossing tunnels: as acts of compassionate conservation 73–74; leveraging existing landscape to benefit multiple species 72–73, **73**
wildlife guards 63–64
wildlife management 44
wildlife mortality 78
Wolvercote Tunnel 80, 83

Yellowstone to Yukon (Y2Y) Conservation Initiative 24–26, 33, 40, 45, 49, 66, 100–101
Zenkewich, Kelly 26

www.ingramcontent.com/pod-product-compliance
Lightning Source LLC
Chambersburg PA
CBHW021145230426
43667CB00005B/264